GW00675457

The Satellite Communication Ground Segment and Earth Station Handbook

For a listing of recent titles in the *Artech House Space Technology and Applications Library*, turn to the back of this book.

The Satellite Communication Ground Segment and Earth Station Handbook

Bruce R. Elbert

Artech House
Boston • London
www.artechhouse.com

Library of Congress Cataloging-in-Publication Data
Elbert, Bruce R.
    The satellite communication ground segment and earth station handbook /
Bruce Elbert.
        p.   cm. — (Artech House space technology and applications library)
    Includes bibliographical references and index.
    ISBN 1-58053-046-X (alk. paper)
    1. Artificial satellites in telecommunication.   I. Title.   II. Series.
TK5104.E45   2000
621.382'5—dc21   00-059390
                                                                       CIP

British Library Cataloguing in Publication Data
Elbert, Bruce R.
    The satellite communication ground segment and Earth station
    handbook. — (Artech House space technology and applications library)
    1. Artificial satellites in telecommunication
    I. Title
    621.3'825

    ISBN 1-58053-046-X

Cover design by Christina Stone

© 2001 ARTECH HOUSE, INC.
685 Canton Street
Norwood, MA 02062

International Standard Book Number: 1-58053-046-X
Library of Congress Catalog Card Number: 00-059390

10 9 8 7 6 5 4 3 2 1

Contents

Preface

Our ability to apply space technology to communication needs on earth is necessarily dependent on the ground segment, and the earth stations that comprise it. Little has been documented on the methodologies and practices that ground segment developers employ to design and implement these facilities. Another aspect is the powerful trend toward consumer-style user terminals that bring satellite communications down to an affordable and practical level. This situation was the primary motivation for this handbook, and we were able to draw on over 30 years of development in this field when putting it together.

The book is organized into 11 chapters that discuss the history, requirements, systems engineering, hardware design, and operations and maintenance of earth stations and user terminals. Our approach is to provide a thorough understanding of the technology and implementation issues for the ground segment, rather than delving into copious detail on each element. This is how we could keep the book to a tractable form and allow the reader to learn how to develop specific approaches for fixed, broadcasting, and mobile satellite applications. A cover-to-cover reading will provide a comprehensive review of almost every aspect of the system, considering real-world aspects of earth station design and engineering. From there, the book provides a reference for subsequent analysis and selection of the appropriate design approach. References containing detailed information on specific engineering issues and solutions are included with each chapter.

We begin in Chapter 1 with a brief history of the satellite communication earth station, which is fundamentally a radio station operating at microwave frequencies. We draw heavily from the fields of radio-astronomy (the first dish antennas were actually radio-telescopes), radar, and terrestrial microwave. Basic engineering principles common to all earth stations are contained in Chapters 2 and 3, the latter delving in particular into the RF link budgeting process for geostationary earth orbit (GEO) and non-GEO satellite transmissions. Chapter 4 addresses ground segment requirements for two-way communications services for telephony and VSAT data applications. One-way (broadcast) service requirements are covered in detail in Chapter 5, is by addressing digital video as well as audio and data broadcasting. The data aspects are of particular interest in the context of applying the Internet Protocol over satellite links.

Chapter 6 addresses what has become a vital concern in designing integrated ground segments for digital services—that of baseband architecture. The topic, which defied proper characterization in the past, involves complex data processing, information transfer over a network, multiple access tradeoffs, modulation and coding, and a host of other aspects of creating an automated network environment that satisfies subscribers and other users. We address an overriding concern about implementing competitive architecture that can make money in commercial services. The more traditional topic of earth station RF and equipment design is addressed in detail in Chapter 7, extending from the general subject of gain budgets (EIRP and G/T) to the specifics of the antennas, high power and low noise amplifiers, and up- and down-converters. Also considered in the chapter are important design and performance issues such as group delay, local oscillator stability, and phase noise. Interaction of all of the RF and baseband elements is covered in Chapter 8, which deals with signal impairments and PC software tools for link analysis and end-to-end simulation.

Our next focus is on fixed and mobile user terminals, covered in Chapter 9 from a design and performance standpoint. Also addressed is the development of baseband and modem functions through the principles of the software-designed radio, as well as specific considerations for handheld satellite phones. Design requirements and implementation of major facilities for large earth stations are covered in detail in Chapter 10. We round out the handbook in Chapter 11 with a comprehensive discussion of operations and maintenance principles based on practical experience with many successful projects over the years. This considers both the operational needs of the earth station and network, as well as lessons learned in managing the human side of the equation.

The handbook was designed for working satellite communications engineers and telecommunications specialists who need to consider how to design, implement, and manage the ground segment. It is also appropriate for specialists in RF, baseband, and digital network technology who wish to gain a better understanding of the total system and the interaction of its key elements. Operations and maintenance personnel should find the book useful as an introduction to ground segments and earth stations, and as a handy reference. From an education standpoint, much of the material was used for an engineering short course at UCLA Extension, where it will become the course text.

There is a lot that goes into a book like this, which captures much of a lifetime's experience with the practical side of applying technology to a real need. I would like to express my appreciation to my technical reviewer, Ray Sperber, for his outstanding guidance and input during development of the manuscript. Also, I wish to thank Bill Bazzy, Mark Walsh, and Barbara Lovenvirth of Artech House for their support from start to finish. I am very appreciative of my wife, Cathy, for helping get the project together and assisting me along way—wherever and whenever she could. Without her, there would be no *Ground Segment and Earth Station Handbook.*

Bruce Elbert
Application Strategy Consulting
bruce@applicationstrategy.com

1

Introduction to the Satellite Communication Ground Segment

Ground segments of satellite communication systems employ a variety of node designs and network configurations in order to provide and manage services delivered to end users. The nodes in these networks range from the large earth stations used as gateways in a telephone network to very small aperture terminals (VSATs) that deliver data communication applications to remote business locations. "Small aperture" in this context refers to a reflector diameter in the range of 60 cm to 2.8m. We must also include low-cost end-user communication devices like desktop and handheld mobile telephones and direct broadcast satellite (DBS) home receivers. This broad range of ground systems and devices employs many of the same hardware and software technologies found in other modern telecommunications and broadcasting networks that are the core of terrestrial wireless and Internet services.

As illustrated in Figure 1.1, the ground segment is that half of a satellite communication system which, quite naturally, resides on the ground. The space segment, consisting of orbiting communication satellites and a satellite control system used to operate and maintain the satellites, is that vital component which relays information between and among the various types of earth stations that make up the ground segment. In a previous work [1], we discuss in detail the design, operation, and management of the space segment, considering the technologies and physical principles that are key to its

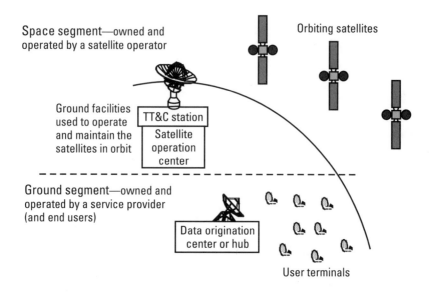

Figure 1.1 Definition of the space segment and the ground segment.

success. In essence, the ground segment is useless without a properly operating space segment. Implementation of the space segment of geostationary earth orbit (GEO) satellites represents many hundreds of millions of dollars; a global nongeostationary (non-GEO) satellite system increases this level to billions. Operators of GEO space segments include INTELSAT, Société Européene des Satéllites (SES), PanAmSat, GE American Communications (Americom), EUTELSAT, Japan Satellite Telecommunications (Jsat), and others. Companies such as Globalstar LLC, ICO Teledesic, and OrbComm operate non-GEO satellite constellations and systems. We will not discuss the space segment in detail in this book, but we will at times need to consider the capabilities and constraints that it imposes as we work to provide the service element of the overall system.

We use the term *earth station* to include classic fixed earth station facilities as well as mobile aeronautical, maritime, and handheld devices. In some cases, earth stations are individually owned and managed (e.g., the teleport operated as a business), but this approach is being overtaken by the integrated total system that provides services to end users. Many readers already own their own user terminals, which are technically small self-contained earth stations. But these terminals cannot operate without the ground segment of the network operator. Examples of the latter include DIRECTV, British Sky Broadcasting (BSB), Measat Broadcasting Network Services

(MBNS), operator of the Astra DBS service in Malaysia, COMSAT Mobile Services, Telia of Sweden, Hong Kong Telecom, PT Indosat, and others.

The first two chapters of this text will lay the groundwork for the detailed discussion of design and operating principles that follows. The intention is to permit readers to devise their own approach to implementing and managing the earth stations and the overall ground segment. The material presented in this book should give readers a head start in the process, allowing them to prepare requirements, perform preliminary analyses, review designs provided by hardware suppliers, and manage the introduction of the overall ground segment. We assume that readers have some level of familiarity with the overall satellite system and the principles of RF engineering and telecommunications networks, particularly those using satellites. We start with a chronological review of earth station technology, beginning in the early twentieth century with the first introduction of radio (synonymous with wireless at that point in time).

The long history of ground segment and earth station development and application dates back to the early 1900s, to the very beginnings of radio communications. Guglielmo Marconi successfully conducted the first transatlantic message transmission in 1901. The transmitters of the day produced a radio signal of around 1 MHz generated from a continuous spark. Radio waves from this frequency and up to about 30 MHz can travel long distances on the ground, since they can be reflected by the ionosphere (instead of an orbiting platform). Later, commercial ships were equipped with these radios so that land-based stations could handle messages sent with the international Morse code. The 1997 movie *Titanic* dramatized how these simple mobile radio stations were the lifelines of communication afforded the greatest ships at sea.

An overall timeline for the evolution of modern ground segments and earth stations is presented in Figure 1.2. It is impossible to present here every significant class of facility and application, since to do so would require an area the size of a movie screen. However, we aim to show each major introduction of ground-based radio technology that contributed to how we use communication satellites in orbit. Other aspects of this history, particularly from the space segment standpoint, can be found in our previous work [1].

1.1 First Space Application of Radio

While HF signals (which lie between 3 and 30 MHz) are much lower in frequency than today's modern satellite links, the principle is nevertheless the same as for the ground segment. Private individuals and experienced radio

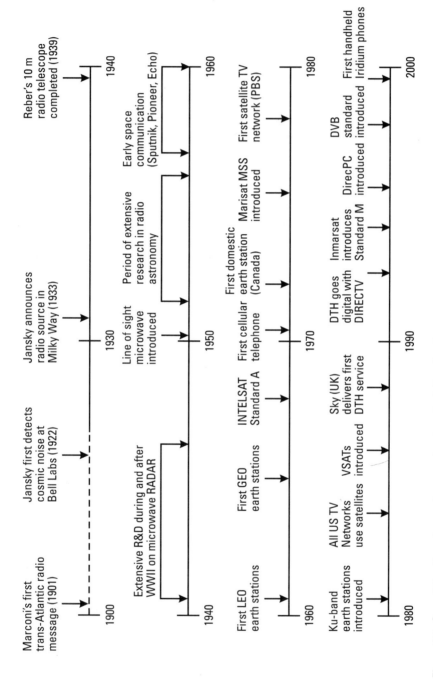

Figure 1.2 Overall timeline of ground segment and earth station development.

hobbyists (called radio amateurs or hams) helped develop radio communications through the medium of amateur radio communication (also called ham radio) and radio astronomy. This author spent many hours as a teenager, college student, and young engineer using the international Morse code to meet fellow amateurs around the world. (International Morse code differs from the original American Morse code because it composes letters with long and short "on" radio transmissions rather than sequences of "clicks" of DC current over telegraph wires.) Starting in the 1920s, amateurs were granted key status in the U.S. domestic regulations and later international regulations as well. They developed their own equipment and antenna systems, and moved the technology from the low frequencies up to the bands close to those of modern satellites.

The first indication of a connection between radio and space came in 1922 when Karl Jansky, a young electrical engineer at Bell Laboratories, was investigating the various sources of noise that hinder long-distance radio communication. While most of the emitters he identified were due to terrestrial sources such as electrical machinery and thunderstorms, there was one unidentified source that appeared and disappeared on a daily basis. The clue as to the celestial nature of this source was the fact that it appeared four minutes earlier every day. From the standpoint of an astronomer, this indicated that the source was extraterrestrial. Karl Jansky's announcement of the source as in the direction of the center of the Milky Way Galaxy was reported on May 5, 1933, on the front page of the *New York Times*.

Most early radio installations looked more like broadcasting sites than earth stations. However, it was an amateur radio operator by the name of Grote Reber who built the first operating parabolic reflector antenna. (The reflector is parabolic in cross section, producing a surface called a paraboloid. However, the term *parabola* is more popular, probably because it is easier to say.) Shown in Figure 1.3 is the 10m (32 ft) wonder he created to explore the radio sky. Reber built this first model himself literally in his own backyard, and made his first discoveries of cosmic radio emissions in 1939. After continued measurement and exploration of celestial radio sources, he published the first survey of the sky at radio wavelengths in 1942 in the *Proceedings of the Institute of Radio Engineers* [2].

1.2 Microwave and Radar Development

Earth stations use microwave frequencies, which lie between approximately 1 and 30 GHz, and therefore owe much to the development of Radio

Figure 1.3 Grote Reber's home-built backyard radio telescope (photo courtesy of NRAO).

Detecting and Ranging (RADAR) systems. Many readers are aware that a radar antenna in the Hawaiian Islands detected imperial Japanese aircraft that bombed Pearl Harbor on December 7, 1941 [3]. However, much earlier, in 1929, G. Ross Kilgore, an engineer at Westinghouse, generated 18 GHz of microwave energy with an experimental split-anode magnetron vacuum tube. This particular device could measure Doppler reflections from moving automobiles and railroad cars. Radar improved in RF power output and sophistication during World War II and shortly thereafter, providing a technology base for terrestrial and satellite microwave communications. Professor Wilmer Barrow of the pioneering Radiation Laboratory at MIT experimented with electromagnetic horn antennas for "static-less" ultrahigh-frequency wave transmission [4]. The system consisted of a conducting tube, a transmitting terminal device, and either a receiving terminal unit or the radiating horn. Other developments at the Rad Lab include a multitude of microwave components like the klystron, numerous waveguide devices like diplexers, and antenna systems for radar and communication applications. An early horn reflector antenna was developed at Bell Laboratories in 1942, a precursor to antennas used in terrestrial and satellite microwave communications. After the war, Raytheon used microwave technology in an innovative (and noncommunication) manner with their invention of the microwave oven. Not surprisingly, the first food item to be cooked was popcorn.

AT&T recognized that microwave technology could increase the capacity and reliability of long-haul communication lines. Line-of-sight microwave links were established across the developed regions of the world during the 1950s and 1960s. The terminal ends and intermediate connection points were very much like earth stations in their design and use, namely to interface the long-distance link with local users. The overall microwave network offers much that a modern ground segment can, although it is tied to the specific routing and associated real estate.

Radio astronomy, while not able to command the investment and revenues of commercial telephone and television services, still benefited from the availability of microwave technology. Here, the challenge is to receive very weak signals (effectively noise within the noise). The parabolic dish antennas grew in size to provide greater ability to discriminate distant radio emitters. The principle behind this is that the width of the narrow antenna beam is inversely proportional to the diameter of the reflector. Thus, radio telescopes in the 30-to-100-m range soon appeared, topped by the giant 305-m Arecibo dish antenna in Puerto Rico (this antenna is actually constructed in a small lake basin and was featured in the James Bond movie *Golden Eye*). Constructed in 1960, and operated by Cornell University under a cooperative agreement with the National Science Foundation (NSF), the Arecibo radio telescope not only receives celestial noise, it can also transmit radar signals to map the planets of our solar system.

Radio telescopes more along the lines of earth stations were constructed at several locations in the United States and around the world. The 91-m (300-ft) Green Bank Telescope was constructed in the 1960s but experienced mechanical failure in 1988. At the time of this writing, another 100-m radio telescope was under construction, this time using the offset parabolic reflector design popular for home DBS installations (and many spacecraft reflector antennas as well). Another type of terrestrial radio communications system is the tropospheric scatter communication link, which employs microwave signals that can be propagated over the horizon (OTH). The tropo installation shown in Figure 1.4 was installed in 1967 by the U.S. Army Signal Corps at Pleiku, Vietnam, where it provided a single-hop link of about 230 km to Nha Trang (four times line-of-sight range). As can be seen, this 15-m antenna points nearly at the horizon to acquire the relatively weak (but reasonably stable) signals that have been dispersed by the troposphere.

Figure 1.4 Tropospheric-scatter terminal in Pleiku, Vietnam (circa 1967).

1.3 First Satellite Earth Stations

Satellite earth stations are different from previous ground-based installations in that they are intended to transmit to and receive from spacecraft. The first of these were used to track the early vehicles that were launched into orbit and deep space. The function is called tracking, telemetry, and command (TT&C) and often includes a requirement to receive various types of sensor information. Such stations could also be equipped for communication with manned spacecraft and orbiting repeaters (e.g., communication satellites). Both the United States and the former Soviet Union introduced these capabilities as part of their respective space programs, which began in the mid-1950s. The first TT&C ground stations were, in fact, radio telescopes that had been modified for bidirectional transmission. An example is the 90-m Goldstone, California, tracking station antenna, which was installed in the late 1950s to track *Explorer 1* by Jet Propulsion Laboratories (JPL), which at the time was under contract to the U.S. Army. In 1959, JPL was transferred to the National Aeronautics and Space Administration (NASA) as part of the still-operating Deep Space Network (DSN).

Also installed at Goldstone was the 30.5-m earth station for use with the Project Echo passive balloon reflector satellite. Bell Labs in Holmdale, New Jersey, provided the other end of the link with their reflector horn antenna [5]. Horn antennas of this type have the added feature of very low side and back lobe radiation and reception, something that helps reduce noise pickup from extraneous sources. Later, this antenna was the instrument used by Bell Labs' scientists A. A. Penzias and R. W. Wilson to make the discovery of the cosmic background noise level, produced by the Big Bang. While experimenting with noise measurements, they were trying to find where about three degrees of excess noise was coming from.

The first active repeater satellite was Bell Laboratories' *Telstar 1*, which allowed the United States and the United Kingdom to communicate real-time TV and voice. As the first true earth stations, these facilities were as large and elaborate as the TT&C stations and large radio telescopes that they emulated. Size did matter because of the low power and antenna gain provided by the little Telstar satellite. Because *Telstar 1* was in low earth orbit (LEO), the earth stations required tracking systems; also, the service was interrupted whenever the satellite was not in simultaneous view of the end points.

All of these earth stations were highly customized and experimental in nature. They were not constructed to provide a service to subscribers and certainly were not operated as a business. In the next section, we consider how earth stations evolved into commercial ground segments installed to provide profitable communications services.

1.4   Commercialization of the Ground Segment

The creation of the Communications Satellite Corporation (COMSAT) and the subsequent launch of Early Bird in 1965 ushered in the era of true commercialization of this medium. It has largely gone unnoticed that one of COMSAT's greatest legacies is its introduction of major earth stations for telephone and TV services to the general public. The very first transmissions over Early Bird used the same earth stations that were constructed for

Telstar; however, these facilities had very limited service capability, both in terms of capacity and physical location. The intention of Early Bird and all of the INTELSAT satellites that followed was to improve the international tele-communications network. This demanded that earth stations be located in essentially every country of the world.

COMSAT worked in cooperation with AT&T, RCA, and others in the United States, and the major post, telegraph, and telephone (PTT) agen-cies of countries around the world, to standardize the design and operation of these large earth stations. They helped found the International Telecom-munications Satellite Organization (INTELSAT), a treaty-based cooperative of national entities around the world. (INTELSAT was created as a quasi-governmental body but spun off its transponder-leasing business to a Netherlands-based satellite operator called NewSkies Satellites.) A typical INTELSAT earth station of the 1960s, such as the COMSAT facility in Etam, West Virginia, represented a substantial investment yet offered con-nectivity with other earth stations through the GEO satellites operated by this consortium. While begun and nurtured by COMSAT, INTELSAT set out on its own in the mid-1970s and initiated a broad range of other services and applications of their space segment. The initial Standard A type of sta-tion (initially requiring 30-m antennas) was joined by more cost-effective Standard B earth stations (at 15m to 20m) that allowed countries to provide domestic satellite communication networks. Similar stations were installed as part of domestic satellite (DOMSAT) systems, such as the Palapa A network in Indonesia (see Figure 1.5).

The INTELSAT system established its preeminence through the 1960s and early 1970s as the number of earth stations grew from hundreds to thou-sands. Using technology and standards originally developed by COMSAT and improved along the way by INTELSAT and its members, the earth sta-tions interoperate effectively. At all times, they maintain quality and order (no small task for a system used by a wide range of operating organizations on every continent).

Headquartered in Washington, D.C., INTELSAT manages the system using its control center, which is connected to TT&C and monitoring sta-tions strategically located around the world. Individual earth stations, which are owned and operated either by members (e.g., the domestic telephone companies or PTTs) or other users who obtain their authority from these entities, are under the direction of INTELSAT's control staff.

While INTELSAT saw its global system grow rapidly in terms of the number of satellites and earth stations, an important new phase of satellite communication appeared in 1971 when Canada launched its first GEO

Figure 1.5 Palapa A light-traffic earth station (10-m antenna) in Indonesia.

DOMSAT, *Anik A* (*Anik* means "brother" in the native Inuit language). Telesat Canada established the first domestic satellite system, which would become a model for more than twenty other countries. It pioneered applications like rural telephone service to remote regions and national TV broadcasting directly to major cities and small communities far from terrestrial transmitters. While INTELSAT required antennas of at least 15m in diameter, the performance of the Anik satellites allowed Telesat to employ 10m and even 4.5m antennas for these services (see Figure 1.6). This innovative system also spurred Canadian industry, allowing several companies to gain a viable foothold as international suppliers of satellite and earth station equipment.

The United States, while a pioneer of commercial satellites and earth stations, took a back seat to Canada and only produced its first domestic network in 1974 with the launch of Western Union's *Westar 1* satellite. Being a purely commercial company, Western Union implemented its ground segment to augment its existing terrestrial microwave network. Westar earth stations were located in major cities around the country to support telephone, telex, and data communications. Later, the Public Broadcasting Service (PBS) became the first U.S. television network to use satellites for program

Figure 1.6 A 4.5-m earth station capable of telephone service (circa 1977).

distribution and backhaul (e.g., the point-to-point transmission of video from temporary sites and remote studios at affiliated TV stations).

Integrated ground segments of the day, primarily used analog technology for telephone, telex, and television service, and depended on human operators to control access and manage services. The first domestic satellite system outside of North America was introduced in Indonesia and its neighboring Southeast Asian neighbors in 1976. The Palapa A satellites (named for a mythical fruit that Gajah Mada, an ancient king of Java, refused to eat until all of Indonesia was united) used the same design as *Anik A*. The ground segment was provided on an integrated basis by an international team consisting of the Indonesian PTT (now PT Telekomunikasi Indonesia), Hughes Space & Communications Company, Ford Aerospace, ITT, and the TRT division of Philips. Internal to every earth station was a single-channel-per-carrier (SCPC) system that provided telephone service on a demand-assigned (DA) basis. This was directly integrated into the Indonesian direct-distance-dialed (DDD) automated telephone network that ITT was installing at the time.

The timeline in Figure 1.2 indicates that 1975 was pivotal for mobile satellite communications, since this was the year *Marisat 1* was launched. This L-band GEO satellite was operated by COMSAT as a means to improve ship-to-shore communications, which at the time was still dependent on ionospheric reflection at high frequency. The first shipboard Marisat terminals still needed a dish type of antenna to provide adequate link performance for SCPC telephone and telex service. At the other end of the link is a land station acting as a gateway to the telephone network in the respective country. This service was so successful that COMSAT created Inmarsat, another alliance to promote the proliferation of earth stations in the global mobile ground segment. We will discuss again how Mobile Satellite Services grew to extend to the air and land, providing communications to a wide variety of user terminals.

1.5 Rapid Developments

The decade of the 1980s saw a big increase in the number of satellites and satellite operators. The GEO began to become crowded in sectors serving North America, Europe, and East Asia. Applications in TV, private data and voice networks, and mobile communications established themselves as the ground segments mushroomed. Private commercial operators such as RCA Americom (now GE Americom), Hughes Communications, and Pan Am

Sat (now combined as a result of a merger in 1997) grew to profitability on the foundation of the cable and broadcast TV industry. Television signals relied on these satellites to serve thousands of locations in North America.

Cable TV systems in local communities created the demand for specialized and high-value programming from HBO, Turner Broadcasting (now part of AOL /Time Warner), Disney, and ESPN. Cable networks numbered more than 100 by 1985 and continue to be delivered through 3-to-5-m C-band dishes at the head ends of local cable TV (CATV) systems. Origination of the network feeds requires extensive studio facilities. Ku-band satellites and ground segments also appeared in the 1980s to take advantage of the smaller dish sizes that this band allows. As a result, VSAT user terminals like that in Figure 1.7 became popular with major retailers like Wal-Mart, Rite Aid, and Kmart, and oil companies like Chevron, Mobile, and Texaco.

The first real DBS system was introduced in Japan in the late 1970s and became very popular by 1985. While offering only two TV channels, the NHK DBS project resulted in several million home DBS installations throughout Japan. The United Kingdom and continental Western Europe saw the explosion of commercial satellite TV during the late 1980s. One could argue that this expansion came about through News Corp's Sky TV and the Astra satellites operated by Société Européene des Satéllites. This powerful combination delivered an attractive programming package directly to subscribers, who bought and installed their own DBS receivers. There is little doubt that this established the viability of direct-to-home (DTH) satellite TV, an application which DIRECTV subsequently took to new heights (but still at GEO altitude).

Inmarsat became the foundation of commercial mobile communications as defense applications of satellite mobile services were already established but with space and ground segments owned and controlled by governments. Several proposals were put forth in the United States and Europe for innovative satellite services using new frequency bands. One was intended for tracking trucks on intercity routes, so that dispatchers knew where their drivers were even if they could not call in. Even though there was a serious shakeout in these early systems, they nevertheless represent an important precursor for the global mobile satellite systems of the next decade.

We cannot forget to mention the cellular telephone networks, which began to reach critical mass during this same decade. While relying completely on local base stations and switching systems, cellular radio proved the value and reliability of automated radio frequency channel assignment, cell-to-cell handover, and intersystem roaming. Similar capabilities were

(a)

(b)

Figure 1.7   Examples of VSAT earth stations for two-way communications: (a) 1.2-m data application (photo courtesy of Hughes Electronics); (b) 1-m video-data terminal (photo courtesy of STM Wireless).

demonstrated on the INTELSAT system with the pioneering SPADE SCPC DA system of the early 1970s and the DA system installed in Indonesia in 1977 for Palapa A. These technologies provided a basis for large global mobile personal communications services (GMPCS) projects like Iridium, Globalstar, and ICO.

1.6   Introduction of Consumer Terminals and Applications

Ground segments that serve individual consumers came into their own during the 1990s. Although many U.S. consumers were enjoying direct reception of cable and TV networks using backyard dishes (typically 2m to 4m in diameter), an explosion of demand for home satellite TV did not really start until DIRECTV was introduced. The space segment for this pioneering system consisted of rather standard Ku-band Broadcasting Satellite Service (BSS) satellites operating in the assigned slot at 101 degrees west longitude (the Astra Fixed Satellite Service, FSS, and BSS satellites had already gained eminence at 19.2 degrees west longitude, based on analog transmission). DTH receivers were installed in American homes by millions of families who either did not have access to cable TV or were unhappy with what the local cable company had to offer.

Complementing the user terminals was the world's largest uplinking center for DIRECTV, constructed in Castle Rock, Colorado. This center was the first to employ MPEG-based video compression and had the unique capability to originate 200 simultaneous channels of programming. The proprietary DIRECTV Satellite System (DSS) demonstrated the technical and operational viability of this type of implementation (as with many U.S. innovations, this proprietary standard has been supplanted in later DTH systems by the European Digital Video Broadcast [DVB] series of open standards). By 2000, approximately 7 million individual subscribers were employing DSS receivers and subscribed to the service (making DIRECTV the United States' third largest cable TV operator after AT&T-TCI and AOL /Time Warner). Echostar, a competitor of DIRECTV, introduced its innovative DISH Network and brought the total U.S. DTH subscriber count to over 10 million, the largest for any country in the world. France, Germany, the United Kingdom, Sweden, Thailand, Malaysia, Japan, Mexico, Brazil, and Argentina are among the nations in which consumers have access to a high degree of choice in TV entertainment. Many of these ground segments use the DVB standard, which contains MPEG-2 video compression and multiplexing, digital audio, encryption and conditional access, and online program guides. DVB provides many standard features that permit set-top boxes to be supplied from multiple vendors, but aspects of the conditional access system are very much closed. All together, 30 million subscribers enjoy this digital DTH satellite service at the time of this writing. This is a boon, not just for the operators but for manufacturers of home receiving antennas and set-top boxes, digital video compression and storage equipment, uplink earth stations, and providers of the programming itself (as well as advertisers) who gain a direct connection to viewers.

Returning to the Mobile Satellite Service (MSS) area, Inmarsat did a good job of bringing satellite communication to the personal level. First with the Inmarsat M and then the M4 (Figure 1.8), it was demonstrated that one could have reliable communication from a highly portable piece of equipment. Likewise, domestic MSS operators Optus Communications (Australia) and American Mobile Satellite Corporation (AMSC) adopted similar

(a)

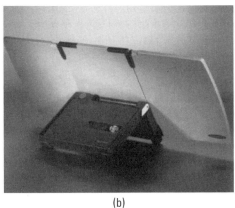

(b)

Figure 1.8 Examples of compact MSS user terminals for use in the Inmarsat system: (a) Standard M; (b) Standard M4 (photos courtesy of NERA).

ground segment equipment, including Improved Multi-band Excitation (IMBE) voice compression technology from Digital Voice Systems Inc. (DVSI). While these terminals were not of the handheld variety, they nevertheless demonstrated that individual calls could be placed and subscribers could be serviced conveniently and relatively inexpensively (e.g., as compared with the best alternative in a remote place, on a ship, or wherever).

The explosive use of the Internet during the 1990s had an impact on satellite communication ground segments. Major corporations that found it an effective way to extend their information technology (IT) systems across a broader business and geographic base adopted TCP/IP and the rest of the Internet suite of protocols. Companies like HNS, Gilat, and STM Wireless enhanced their VSAT offerings by better supporting TCP/IP. Oddly, much misinformation was propagated during the 1990s that TCP/IP was incompatible with GEO satellite links due to the combination of propagating delay and noise-induced errors. The leading suppliers of VSAT terminals and the associated hub stations devised effective solutions through protocol "spoofing" (a technique also used by router vendors Nortel and Cisco to improve performance) to allow satellite networks to challenge terrestrial solutions. HNS also introduced the first personal satellite Internet solution, DirecPC, to allow subscribers to reach 400 kbps downlink speeds. This service was available in 12 countries at the time of this writing.

1.7   Ground Segments in the Twenty-First Century

The year 1999 was a watershed for satellite communications because of the commercial introduction of GMPCS through the Iridium and Globalstar systems. With a space segment mirroring a typical cellular network, Iridium provides the infrastructure for fast, flexible mobile communications with and between any place on or near planet Earth. The first Iridium handheld telephone, illustrated in Figure 1.9, is not small enough to put in a shirt pocket or small purse, but its communication capability is no less impressive (however, the pocket-sized Iridium pager is still a viable alternative). The ground segment itself provides the landing point for terrestrial phone calls and allows this operator to manage the system. Close behind Iridium and Globalstar is ICO, which provides telephone service to handheld devices of similar characteristics. At the time of this writing, Iridium was in bankruptcy and Motorola, operator of the system, was in the process of planning its closedown. The new owner of ICO, Craig McCaw, announced that he would

Figure 1.9 First-generation MSS handheld mobile phone for the Iridium system (photo courtesy of Kyocera).

redefine the service to focus away from handheld telephone service and toward medium-rate wireless data applications.

A great deal of excitement exists about advanced satellite systems that will offer interactive broadband services to homes and business. Unlike DBS and VSATs, the coming generation of broadband systems intends to compete with the fiber optic and cable networks in the provision of high-speed Internet and multimedia services. Two non-GEO systems—Teledesic and SkyBridge—have acquired frequency spectrum to support their very large projects. At the time of this writing, SkyBridge had backing from Alcatel and was making deals with Boeing and others to round out its program. However, little was known about the nature of the ground segment and user terminals for either system.

Ground segments and earth stations have at their foundation the very beginnings of radio and microwave. Through innovations in digital signal processing, RF electronic design and manufacture, and transmission innovations in forward error correction and compression, earth stations have been driven from being major installations (and national monuments in some countries) to home electronic units costing $100 or less. What can we expect in the next hundred years?

This, of course, is a difficult question to answer. But one observation is clear—the earth stations and user terminals to be developed and employed around the world will be responsive to the business demands and new applications that address user needs on a competitive basis with terrestrial alternatives. Previous obstacles such as the difficulty of owning and operating earth stations and their high cost are behind us. Now, it is a matter of developing the right infrastructure design and making it available at the right time.

References

[1]   Elbert, Bruce R., *Introduction to Satellite Communication*, 2d ed., Norwood, MA: Artech House, 1999.

[2]   Reber, Grote, "Cosmic Noise," *Proceedings of the Institute of Radio Engineers*, Vol. 30, pp. 367.

[3]   Strull, Gene, "Westinghouse Microwave Systems and Technology," *1998 IEEE MIT-S International Microwave Symposium Digest*, Vol. 2, Baltimore, MD, June 7–12, 1998.

[4]   The Rad Lab's Microwave Traditions at RLE, RLE Currents, Vol. 4, No. 2, Spring 1991 (http://rleweb.mit.edu/radlab/radlab.HTM).

[5]   http://www.bell-labs.com/history/

2

Earth Station Design Philosophy

The previous chapter provided a historical perspective for the ground segment, laying out how earth stations were created and evolved into higher forms. As a radio communication facility, an earth station receives and, in many cases, transmits a properly formatted signal on a reliable and affordable basis. The first earth stations were designed as major facilities that could house the necessary electronic equipment. Like the radio telescopes and tropospheric scatter sites discussed in Chapter 1, these earth stations were impressive in their scale. They bear some resemblance to major earth stations in modern networks used as uplinks, concentration points, and network management centers. Subscriber terminals, on the other hand, must have fewer components and be simple to operate and maintain. A single-function design philosophy was pioneered with C-band backyard dish receivers and the first VSATs drawn from the consumer electronics and telecommunications equipment businesses. We next review basic earth stations and their functions as a foundation for the detailed design discussions in subsequent chapters.

2.1 The Major Earth Station

As the communication hub of literally every application system, major earth stations come in varying sizes and configurations to satisfy the requirements in one or more of the following functions:

- Telephone or telecommunication gateway, allowing remote user terminals or other gateways to gain access to public or private terrestrial networks including the PSTN and the Internet;

- Broadcasting uplink to originate video programming, audio (radio) programming, and data and other noninteractive forms of information;

- Hub in a star network, allowing remote user terminals to connect back to a central location to access a host computer, servers, telephone switching equipment, and private video transmission;

- Network control center, to process requests from remote terminals for service and satellite bandwidth, and to manage the overall satellite telecommunication network.

Because of the modulator and the flexible nature of any large earth station, we can use the generic block diagram provided in Figure 2.1 to explore the various subsystems. Probably the most essential and sensitive aspect is the RF terminal (RFT), which provides the critical direct interface to the space link. The RFT radiates in the assigned frequency band, under control or

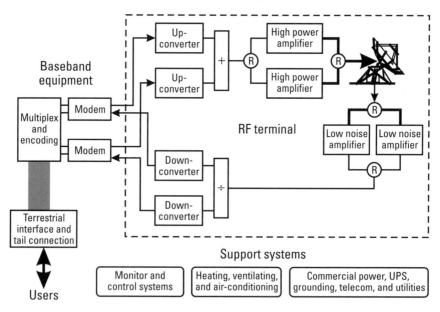

Figure 2.1 Operating elements of a major earth station, including RF terminal, baseband equipment, terrestrial interface and tail connection, and support systems.

direction of the satellite or ground segment operator. Connection to base-band equipment is at a standard intermediate frequency (IF), to simplify connecting different types of transmission equipment, whether analog or digital. The IF center frequency itself is determined by the RF bandwidth of the carrier on the link to the satellite. The actual RF frequency is established by the up converter and down converter, for transmit and receive, respectively. Typical IFs include:

- 70 MHz, supporting a usable bandwidth of 36 MHz (the typical transponder bandwidth at C-band) up to a maximum of 54 MHz (one of a number of bandwidths used at Ku-band);
- 140 MHz, supporting wider IF bandwidths, such as 110 MHz (the upper end of what has been used on some Ku-band satellites);
- 1550 MHz, allowing a total bandwidth of 500 to 1,200 MHz to be passed from the low noise block (LNB) converter on the antenna to internal electronics of the earth station.

The most visible element of an earth station is the outdoor antenna, which can take many forms and sizes to correspond to the application requirements illustrations (gain, beamwidth, and isolation). For GEO satel-lites, this antenna can remain essentially fixed on the satellite, moving only for initial alignment and if a different orbit position must be viewed. Non-GEO systems put much tougher requirements on this class of antenna to direct the beam at moving spacecraft. Satellites in inclined geosynchronous orbits place a requirement on the antenna for slow-rate tracking. Some or all of the remaining electronic equipment of the RFT is mounted to the antenna to reduce losses and, in the case of reception, reduce the noise temperature. All of these concepts are reviewed below and covered in detail in Chapter 3.

The electronic equipment of the RFT has not changed much in princi-ple since the first systems were constructed. This is because all must perform the basic functions of translating the IF signal to the operating RF frequency, amplifying it to an adequate power level, and connecting it to the antenna (e.g., for the uplink). For reception (the downlink), the RFT collects and amplifies the signal, and then translates it in frequency to IF. Requirements for each of these components are determined through a detailed budgeting process, where the overall specifications of the earth station are allocated to components.

The elements of a compact RFT can be attached to the antenna struc-ture (possibly the feed itself). In large installations with many high-power amplifiers and redundant systems, the RFT may be contained in its own shelter by the antenna, or installed in the earth station building itself. The latter reduces RF equipment maintenance difficulties but introduces more transmit waveguide loss. An exception might be where the antenna is mounted on the roof immediately above the HPAS.

While the RF terminal design is dictated by the satellite link requirements, the baseband equipment is highly specialized and customized for the particular multiple access (ma) method and end-user service requirements. Because baseband equipment is composed of hardware elements performing signal encoding, multiplexing, and modulation, as well as the reverse of these functions, it is usually supplied as an integrated system. In recent years, much of the complex functionality is provided by a custom software component that runs on a dedicated computer or other processor. We find the greatest complexity, and along with it, flexibility, in time division multiple access (TDMA) baseband systems. Frequency division multiple access (FDMA) is usually simpler to configure and manage, since transmissions are kept apart in frequency. An exception to this rule is demand assignment (DA), wherein the channels are assigned dynamically on a call-by-call basis. In this instance, the baseband system requires a considerable degree of automation under software control. The remaining MA method, namely code division multiple access (CDMA), is likewise complex and specialized, with baseband highly customized as well.

The major earth station is in effect a production facility that must operate and be maintained by staff on a 24-hour-a-day basis. Necessary support functions for the generic earth station are indicated at the bottom of Figure 2.1 and summarized as follows:

- *Monitor and control (M&C) systems.* Earth stations can be operated locally by technical staff or, if the system allows it, remotely from a control center. A properly designed M&C system allows staff members to detect, troubleshoot, and resolve technical problems in a timely manner. Most allow operators to configure portions of the station for service and to change the functions being used by customers and subscribers. The facilities to do this are integrated into the equipment and overall station and some could be remoted to a control center by a specialized network.

- *Heating, ventilating, and air-conditioning (HVAC).* Much of the indoor equipment is similar in design and construction to high-quality computer systems, memory, and peripherals, and therefore should be kept in a hospitable environment. This consists of maintaining temperature and humidity within the proper range for correct and long-life operation. Typical specifications are as follows:

- Temperature: 15° to 25° C
- Humidity: 30% to 70%
- Dust: A consideration in particularly dusty areas. Air should be filtered for particulate matter

- *Power and utilities.* Commercial prime power is typically rated according to the national standard in the particular country. More critical is the tolerance on this voltage, as some equipment might not have been provided with adequate power regulation.

- *Emergency safeguards* (fire, flood, earthquake, heavy wind). Potentially the most difficult and costly factors to address adequately. See applicable sections in Chapter 10.

2.2 User Terminals

While major earth stations have many of the same physical and electronic characteristics, such is not the case with a user terminal (UT), which is intended for a specific application. This is clearly illustrated by comparing a DBS home receiving system to a handheld GMPCS mobile telephone, which is like comparing a VCR to a cordless phone. What is important is that these devices must be simple to operate, reliable in service and function, attractive in appearance and design, and affordable in cost (all of this, of course, is relative). The UT is designed and manufactured like other consumer electronic products, using very large scale integration (VLSI), application-specific integrated circuits (ASIC), and special software programming in read-only memory (ROM). Also, advances in digital signal processing allows designers to convert complex modulation schemes and other algorithms into DSP implementations, which are repeatable in performance and much more cost effective than former analog designs. With manufacturing volumes now reaching hundreds of thousands, or even millions of units, it makes sense to use the latest technology and production systems. Considering first the typical VSAT, its RFT is provided as an integrated package of antenna, feed, and transmitter/receiver (transceiver). On the other end of the interfacility link (IFL) coaxial cable, which carries the IF signals and power for the RFT, is the indoor unit that contains the baseband and interface equipment. The indoor equipment of a conventional VSAT is functionally similar to that of the larger type of earth station used as the hub in this type of star network. However, the quantity of equipment, its bandwidth capability, and degree of redundancy are much less than what one would encounter in the typical large earth station. The indoor unit is contained in an enclosure about the size of a standard PC. Like a PC, it is possible to configure the indoor unit for a particular set of services and transmission features (such as protocol operation, MA, error correction, and control of channel assignment).

We could compare the DTV TV receiver to the common VCR found in most homes. A 45-to-60-cm offset-fed parabolic antenna is used to capture the signals, which are amplified and block translated to a 1 GHz IF in the LNB. The coaxial cable carries about 500 MHz of spectrum into the home to make all of the transponder channels available for demodulation. DC current to power the LNB is carried back over this cable. Within the set-top box, another down converter is tuned to the frequency of the carrier where the desired video channel is located. This assumes a multiple-channel-per-carrier (MCPC) time division multiplex (TDM) scheme such as used in DVD and DSS. The carrier is demodulated down to a single bit stream allowing errors to be removed through forward error correction (FEC). After demultiplexing, the nearly error-free MPEG data is converted back to the appropriate analog video format with stereo sound and other ancillary data (e.g., the electronic program guide and conditional access). Beginning in 2000, DTH receivers were appearing that also include an internal hard drive and control circuitry to allow watchers to pause, replay, and record programs for later viewing.

We complete our brief discussion of user terminals with the handheld GMPCS class of mobile telephone. There are some similarities with VSATs and DTH receivers, but the new demands for compactness require greater customization and miniaturization. Another consideration is battery operation for an extended period (e.g., 2 to 4 hours talk time, 24 to 48 hours standby). Many of the elements in the block diagram are familiar: transmit and receive operation as in a VSAT, inclusion of a digital modem, encryption, forward error correction and microprocessor control as in a DTH receiver, and simplified functional design as in both.

2.3 Design Principles

Earth stations and the ground segments that they comprise are created according to engineering principles established over the twentieth century. This gives us an excellent base with which to understand how these facilities and devices can be built and operated efficiently and cost effectively. The following discussion is an introduction to the topic, to be expanded greatly in coming chapters. We review the basic physical principles of microwave design, satellite and telecommunications systems engineering, and systems operations and maintenance. Readers wishing more fundamental background information may refer to the references at the conclusion of this chapter [1, 2, 3].

2.3.1 Microwave Systems Engineering

Communication between the RFT and the satellite is governed by the basic principles of electromagnetic wave propagation. This spectrum of radiation covers everything from AM radio to light, but we are interested in microwave frequencies between about 1 and 50 GHz (the segment above 30 GHz is more aptly called millimeterwave). As Figure 2.2 indicates, the microwave spectrum is broken up into the familiar L, S, C, X, Ku, and Ka bands actively used in commercial and military satellite communications. These are applied to the different services, namely:

- *Fixed Satellite Service* (FSS), intended for communication among fixed locations on the earth by public telecommunication operators (direct reception by the public is not intended, but is employed throughout the world as another application of this band). Services tend to be broadband in nature (e.g., greater than 100 kbps and typically in the range of 1 to 200 Mbps) due to the RF bandwidth available and the link performance of fixed directional antennas on the ground. Originally allocated for GEO satellites, at least one non-GEO constellation has been granted noninterfering access to this spectrum through subdivision of the allocation.

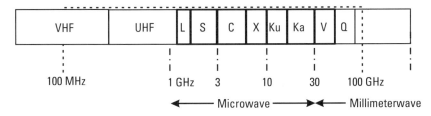

Figure 2.2 Designation of microwave and millimeterwave bands using letter abbreviations (scale is logarithmic and boundaries are approximate).

- *Broadcasting Satellite Service* (BSS), the bands intended for direct reception of broadband information by the public. The Ku-band BSS segment has been assigned into channels and orbit positions for use by individual nations according to a predetermined plan. This service is fundamentally reserved for GEO satellites, although entry is being allowed for at least one non-GEO satellite system within this same spectrum.

- *Mobile Satellite Service* (MSS), the bands around L and S band, which are available for communication with mobile earth stations, including ships, aircraft, vehicles, and persons. The Inmarsat system was established at L-band with GEO satellites followed by domestic satellites for land-mobile services. The public later assigned L and S bands to non-GEO satellite networks for GMPCS applications, although these were slow to gain general acceptance at the time of this writing.

Since many readers are familiar with the basic property of electromagnetic wave propagation in free space, the following is a summary of radio engineering principles, demonstrating their simplicity in mathematical terms. The key elements in the associated RF link are shown in Figure 2.3. For the

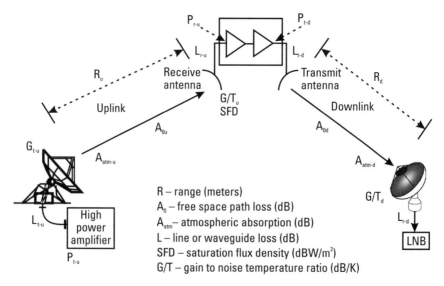

Figure 2.3 Key elements and terms for the uplink and downlink.

typical microwave link on a space-to-earth (downlink) or earth-to-space (uplink) path, the free space loss can be expressed in dB as:

$$A_0 = 183.5 + 20 \log F + 20 \log (R/35,788)$$

where F is the frequency in GHz, and R is the range in km.

The factor 20 in front of the log provides the squaring of both the frequency and range in the equation, and the denominator of R (e.g., 35,788) is the mean altitude of GEO. This formula, however, adjusts automatically for any range, including that of a LEO system at, say, 1,000 km.

Another important principle is that the performance of the microwave link can be predicted using the power balance equation, namely:

$$P_r = P_t - L_t + G_t - A_0 - A_{atm} + G_r - L_r$$

where P_r is the power reaching the receiver; P_t is the transmitted power; L_t is the waveguide loss between the transmitter and the transmitting antenna; G_t is the gain of the transmitting antenna; A_0 is the free space loss; A_{atm} is the sum of all atmospheric losses; G_r is the gain of the receiving antenna; and L_r is the waveguide loss between the receiving antenna and the receiver. The receiver is the first low noise amplification stage of the earth station or satellite, depending on whether this is the downlink or uplink, respectively. Transmitted power, P_t, is provided by a high power amplifier (HPA) within the sending end of the link. The most important single element on each end, other than the LNA or HPA, is the antenna used to either radiate the signal into space or to capture it on the receive side. As readers are aware, the performance of the antenna in terms of gain and beamwidth is governed by its effective area (the physical area of the antenna, adjusted downward by the aperture efficiency).

For the typical parabolic reflector type of antenna illuminated by a feedhorn of some type, the on-axis gain of the main beam can be calculated (as a ratio) from [4]:

$$G = \eta \, \frac{4\pi A}{\lambda^2}$$

where η is the aperture efficiency (λ), A is the physical area of the aperture in square meters, and λ is the wavelength in meters. We can express this formula for a parabolic reflector antenna in a convenient form as follows:

$$G = 10\log(110\eta F^2 D^2)$$

where G is the gain in dBi (e.g., gain relative to an isotropic radiator of 0 dB gain), F is the frequency in GHz, and D is the physical diameter in meters. This assumes a center-fed circular reflector antenna. For noncircular offset-fed reflector antennas (discussed in Chapter 3), it is a common practice to use a mean dimension and adjust for error in the value of η. An important parameter for a parabolic reflector antenna is the half-power (3 dB) beamwidth [5]:

$$\theta_{3dB} \approx \frac{70\lambda}{D}$$

or equivalently for F in GHz and D in meters:

$$\theta_{3dB} \approx \frac{21}{FD}$$

Proper operation of the antenna in the system depends on the polarization of the electric component of the wave, which can be either linear or circular. Since most systems employ frequency reuse in the same beam, it is vital to provide adequate polarization isolation, measured as the difference in dB between the desired and undesired polarizations. For linear polarization, this amounts to proper rotational alignment of the feedhorn, while for circular it is inherent in the feedhorn design itself. The other important characteristic with respect to interference is sidelobe isolation. The following practical formula has become a specification for the maximum expected sidelobe level, in dBi:

$$G(\Theta) = 29 - 25\log(\Theta)$$

where Θ is the angle measured between the main beam and the sidelobe direction. This formula was originally adopted by the U.S. Federal Communications Commission and has become a global standard for frequency coordination between GEO satellite networks. A version of the formula originally adopted by the International Telecommunication Union (ITU) permits 3 dB greater sidelobe level:

$$G(\Theta) = 32 - 25\log(\Theta)$$

The FCC version provides greater confidence of satisfactory operation and would allow closer spacing of satellites as well (in fact, it was adopted back when the FCC wanted to reduce orbit spacing from 4 to 2 degrees).

The effective isotropic radiated power (EIRP) and the gain-to-noise-temperature ratio (G/T) are important figures of merit for earth stations as they establish the RF link performance. The transmit parameter, *EIRP*, is composed of the first three variables in the power balance equation:

$$EIRP = P_t - L_t + G_t$$

expressed in dBW (i.e., dB relative to one watt). From this relationship we see that EIRP can be improved by increasing either P_t or G_t, or by reducing the loss L_t.

The earth station *G/T* is the other figure of merit, i.e.,

$$G/T = 10 \log (G_r - L_r - T_{sys}), \text{ in dB/K}$$

where T_{sys} is the combined receiving system noise temperature in Kelvin. *G/T* has the curious units of dB/K. To compute T_{sys}, we must know the noise temperature of the earth station receiver (assuming this is the downlink) along with the losses associated with the antenna. The following is the basic formula for computing the system noise temperature:

$$T_{sys} = T_a/l_r + (1 - 1/l_r) \bullet 290 + T_{re}$$

where T_a is the antenna temperature (e.g., the cosmic and background noise picked up by the antenna feed and reflector), l_r is the receive waveguide loss expressed as a ratio greater than 1 (e.g., $l_r = 10^{Lr/10}$), and T_{re} is the receiver equivalent noise temperature, e.g., the familiar noise temperature rating of the LNA or LNB, as appropriate.

The purpose of using *G/T* as a figure of merit for an earth station is that we can directly apply it to the link budget to provide a measure of carrier to noise. We see that

$$C/T = EIRP - A_0 - A_{atm} + G/T$$

where *C/T* is the carrier-to-noise-temperature ratio, in dBW/K. Noise power density (N_0 in watts per hertz of bandwidth) is proportional to noise temperature, with the proportionality constant being Boltzmann's constant. We then have the following relationship for carrier to noise density as a ratio:

$$C/N_0 = C/kT$$

where k is Boltzmann's constant (1.380622×10^6 watts/Hz-K).

C/T is converted into carrier-to-noise ratio (C/N) by taking into account the bandwidth of the RF signal that the link supports. The following formula is expressed in dB:

$$C/N = C/T - 10 \log k - 10 \log B$$

where k is Botzmann's constant (note that $10 \log k = -228.6$ dBW/K/Hz) and B is the carrier occupied bandwidth in Hz which is determined by the data rate in bps, R_d, and type of modulation (i.e., $B \approx 0.6 \bullet R_d$ for quadrature phase shift keying [QPSK]).

Because the atmosphere is complex, A_{atm} is really a combination of several losses caused by individual constituents (e.g., oxygen, nitrogen, water vapor, and rain). Specific losses due to air and rain are illustrated in Figures 2.4 and 2.5, respectively [5]. We see that clear air loss is generally well under 1 dB while rain attenuation above about 12 GHz ranges from 2 dB to over 20 dB, depending on the frequency and rain rate. Rain rate, in turn, is a statistical factor produced by the local climate; e.g., we can expect intense heavy rain (resulting in extreme rain attenuation at peak times) in tropical climates like Java, and little rain attenuation during dry months in arid climates like California.

2.3.2   Modem Design

The fact that digital communication has taken over satellites as the primary mode of information transmission is not surprising, since many important innovations in digital communications, such as TDMA and digital voice compression, were first applied to satellite links [6, 7]. As the title of this section suggests, the key earth station component for sending and receiving data is the modem. Readers are familiar with voice-band modems used to send and receive faxes and to connect to the Internet over the public switched telephone network (PSTN). This term is a contraction of modulator/demodulator, reflecting the fact that it provides the dual function of converting a data stream into a modulated carrier, and vice versa. For satellite communication in particular, this is critical because the modem must optimize the transition

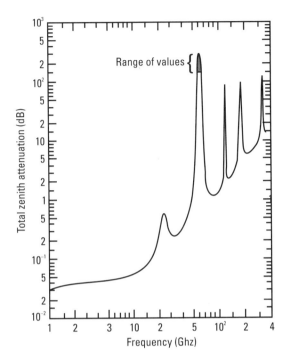

Figure 2.4 Typical atmospheric absorption (dB).

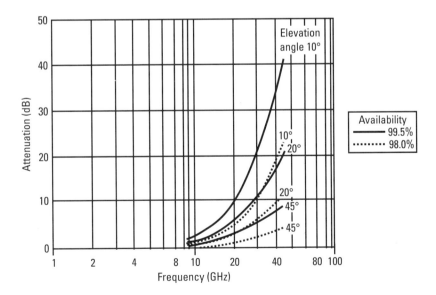

Figure 2.5 Rain attenuation versus frequency for 99.5% and 98% availability in a temperate climate.

between digital information and the analog carrier. The digital side interfaces with the data source after any preparatory processes like multiplexing, compression, and encryption. Often, the modem contains forward error correction (FEC) to reduce the required C/N for satisfactory bit error rate (BER) after the satellite link. The analog functions, to be described below, shape digital pulses to reduce the RF bandwidth, modulate the pulses onto the carrier, and in the case of higher order modulation like QPSK, 8 phase PSK, and 16 quadrature amplitude modulation (QAM) transform the waveform into signals that conserve bandwidth.

The data stream and associated modulated carrier for a biphase shift keying (BPSK) modem are shown in Figure 2.6 [8]. We see a 180-degree phase reversal whenever there is a transition from a zero to a one (or vice versa). Because this illustrates BPSK, the signal does not require the receiver to have an absolute phase reference in order to resolve the sense of the received bit. A block diagram of the modulator is provided in Figure 2.7 to show how nonreturn to zero (NRZ) data is modulated onto the sinusoidal carrier by a balance mixer, then amplified and bandwidth filtered to reduce the spectrum width. The last step is needed to suppress sideband energy and thereby control adjacent channel interference (ACI). Power amplifiers and other nonlinear devices positioned after this filter may re-create the

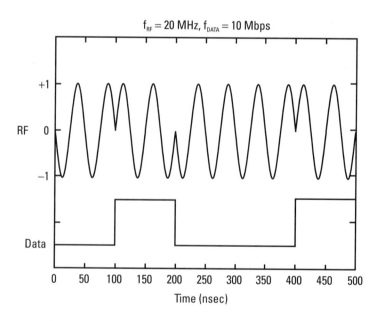

Figure 2.6 BPSK time domain waveforms.

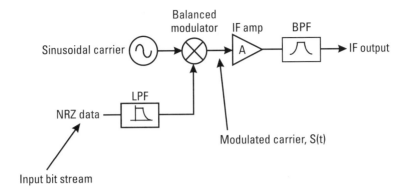

Figure 2.7 Simplified BPSK modulator block diagram.

sidebands and produce ACI in spite of any filtering at the input to the demodulator. The demodulator section (Figure 2.8) determines the final performance BER, which is the ratio of incorrectly received bits divided by the total received bits during a given time interval. These errors are introduced when the received bandwidth contains instantaneous peaks of noise and interference that cancel the desired signal. The problem is further complicated by distortion of the bit pattern itself by bandwidth limiting (which is necessary to reduce the total noise power) and channel impairments such as group delay and AM to PM distortion.

Digital communication link performance is largely determined by the ratio of signal to noise, measured by a parameter called the energy-per-bit-to-noise-density ratio (E_b/N_0):

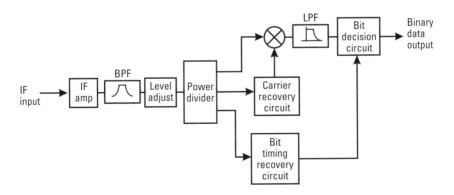

Figure 2.8 Simplified BPSK demodulator block diagram.

$$E_b/N_0 = C/N + 10 \log (B/R_b)$$

The factor B/R_b corrects for a bandwidth offset, namely that B is two sided and R_b is one sided. In a crude sense, E_b/N_0 is 3 dB greater than C/N, although the precise adjustment must account for other factors such as the number of levels (e.g., 4, 8, or 16) per symbol.

A theoretical plot of bit error rate (BER) versus E_b/N_0 is provided in Figure 2.9 for a typical modem performance curve under thermal noise (but excluding other factors such as adjacent channel interference (ACI) and intermodulation distortion (IMD)). Further improvement would be obtained from FEC as this reduces BER by several orders of magnitude at the expense of increasing bandwidth occupancy. However, this tradeoff is almost always favorable, because BER can be improved more effectively by adding FEC than by increasing EIRP.

2.3.3 Multiple-Access Control

Much attention in recent years has been focused on MA systems and the corresponding benefits to the technical performance of the system and ground

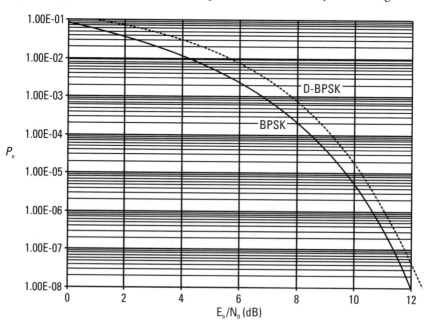

Figure 2.9 Theoretical probability of bit error (P_e) for BPSK and differential BPSK.

segment. The following, in conjunction with Figure 2.10, summarizes the three primary alternatives for multiple access:

- *Frequency division multiple access* (FDMA), where individual earth stations separate their transmissions from each other by uplinking them on different frequencies. This is the simplest MA technique, since stations transmit to the satellite without coordination and with minimal interaction. Single channel per carrier (SCPC) is that form of FDMA where each individual signal (voice conversation, TV program channel, or data stream) gets its own carrier within the satellite repeater. The alternative is to multiplex several channels into a carrier's baseband, which is called multiple channel per carrier (MCPC). In either case, loading of the transponder on a bent-pipe satellite repeater requires management of multiple carriers and the resulting RF intermodulation distortion (IMD). Due to the required amplifier output backoff of 3 to 5 dB, the capacity of the transponder is reduced by at least 50%.

- *Time division multiple access* (TDMA), where separation is achieved by having earth stations transmit their data as bursts at different times, according to a preset time frame. Thus, the transmissions

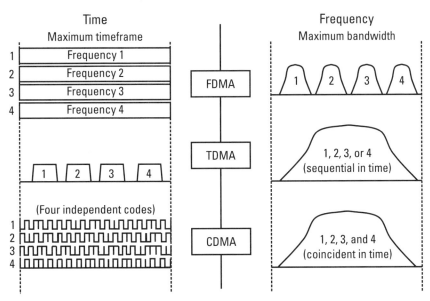

Figure 2.10 Time and spectrum illustrations of FDMA, TDMA, and CDMA, assuming four transmitting earth stations.

must be synchronized in time to prevent collisions among the transmissions when received at the satellite. An alternative form of TDMA, called ALOHA, allows earth station transmissions to be uncoordinated and so introduces the possibility of collisions and a corresponding requirement for automatic retransmission. In wideband TDMA, the burst transmissions at between 60 and 250 Mbps use the full bandwidth and power of the transponder, resulting in nearly 95% efficiency (allowing for the necessary synchronization overhead and guard time between bursts). Alternatively, TDMA networks can use lower data rates (between 64 Kbps and 15 Mbps) to share the capacity of a transponder in an FDMA mode and reduce the uplink power required from the earth station. The inherent digital feature of TDMA has made it the most popular multiple-access technique for VSAT networks.

- *Code division multiple access* (CDMA), where earth station transmissions are encoded using direct sequence spread-spectrum waveform. This is another popular technique obtained by mixing the user data with a very high-speed stream of bits from a pseudo-random noise (PN) generator. Several carriers may be transmitted on the same frequency but are separated by virtue of the different spreading codes. The information on any particular CDMA channel is recovered at the receiver by multiplying the incoming PN-modulated data by the original PN stream. Prior to data recovery, the CDMA receiver must synchronize to the spreading sequence and lock onto its precise timing (a technique called autocorrelation). CDMA has been made immensely popular by the success of IS-95, the digital cellular radio standard from Qualcomm that is based on this multiple-access mode [9].

The baseband characteristics, equipment configurations, software requirements, and management systems for these three MA techniques are very different. What is more, the particular design for a given supplier will likely be incompatible with that of another. An exception is for INTELSAT, Inmarsat, and EUTELSAT standards that have open architectures intended for use by multiple international operators. It would be necessary to examine, in detail, the corresponding elements of the earth station to uncover how the MA functions have been implemented. In simplified block diagrams, MA shows up as a box or a footnote, but in reality, the necessary logic can be dispersed throughout several elements. For example, the modem would appear as a single box in the diagram, but will differ in its internal design.

Shown in Figure 2.10 are the time and spectrum diagrams for the three MA methods, indicating how one transponder would be occupied by four earth station transmissions. In the FDMA time frame, all four transmissions are visible, with each at an independent frequency. There are three potential concerns with FDMA: IMD produced by a common satellite or earth station RF amplifier, ACI due to unfiltered out-of-band spectrum energy or error of the center frequencies (e.g., frequency tolerance), and transponder overdrive due to carrier power imbalance.

In wideband TDMA, the transmissions occur at different times but employ the full transponder bandwidth; separation is guaranteed by proper timing of the bursts and adequate guard times to prevent overlap. The fact that only one wideband spectrum is shown is not a concern, because the stations do not transmit at the same time (the spectrum illustration is, in effect, a snapshot taken when only one of the four stations is transmitting). A picture of narrowband TDMA would look like the FDMA case, since the burst occupies only a fraction of the total bandwidth. However, the picture is once again a snapshot in time, with several earth stations sharing the narrowband RF channel.

For the CDMA spectrum arrangement in Figure 2.10, the summation of signals from the four stations is because they transmit at the same time and may each utilize the full bandwidth. The way that the signals are separated at receivers is through the autocorrelation function previously mentioned. To provide independence, CDMA requires that different PN codes be used by each transmitting earth station; otherwise the receiver will not be able to separate the data. In practice, there is a limit to the number of simultaneous CDMA signals on the same frequency because unwanted transmissions appear as additional receiver noise. Also, RF power level in the satellite repeater and earth station receiver must be kept within a narrow range to prevent elevated carriers from producing more RF interference than can be tolerated on a system basis. CDMA has an advantage over TDMA and FDMA in that it can reject narrowband RFI that could appear in the occupied bandwidth.

The control of the earth station transmissions in a common MA ground segment is critical to the overall management of space segment resources and the delivery of effective user services. In FDMA, this management can be manual, using centralized network control, where operators can monitor transmissions and react quickly to problems. An automated demand assignment multiple access (DAMA) system would be under computer control and comparable to a first-generation analog cellular radio network—i.e., carrier frequencies are assigned temporarily for connections between pairs of earth stations, and then taken back for use by other stations when needed.

TDMA is the embodiment of a digital network via satellite. Also used in the digital advanced mobile phone system (D-AMPS) and global system for mobile communicaitons (GSM) cellular radio standards, this MA technique was originally pioneered on satellite for the INTELSAT system [6] [10]. The principle benefit is maximum usage of the available power and bandwidth without experiencing IMD. The control of burst transmissions within the time frame is provided through a synchronization system and traffic-control methodology exercised by a central network management facility (located at the hub station or one of the earth stations in the network). Control in a CDMA network is simplified in light of the fact that neither network frequency nor burst timing is critical to operation. However, there are still issues regarding the total loading of the transponder and control of individual power levels. These networks are designed with automatic schemes to adjust parameters dynamically, in response to traffic and power loading.

2.3.4   End-to-End Satellite Networks

The ground segment in satellite communication is not an end in itself, but rather is a piece of the overall delivery system for telecommunication or information services. Therefore, we must keep in mind that our satellite network does not stand on its own, and cannot operate in a vacuum (unlike the space segment). As shown in Figure 2.1, the typical earth station interfaces with a terrestrial segment in order to connect to the final user or, in many cases, an existing public network. The best example is a telephone gateway in an MSS network, where the satellite network connects calls that originated in the public switched telephone network (PSTN).

In many satellite applications, the user terminal is self-contained and does not need to interface with anything or anyone other than the end user. An example is an MSS user terminal in the form of a radio telephone instrument (e.g., a satellite cellphone). As long as the user understands how to operate the device, there is no additional end-to-end interface requirement. The other end of the communication link may still have to be transferred to the PSTN, in which case the MSS service must properly interoperate with existing public services. This can be an extremely difficult task, because of the multitude of operating conditions, types of calls, and differences that exist among forms of the PSTN in different countries (and sometimes within the same country).

There may still be interface requirements at the user terminal when the service must be connected to another device such as a PC or TV set. The PC must have the appropriate interface connector, signaling, and software to assure that the service works correctly. We mentioned previously that data networks using TCP/IP usually require some form of protocol spoofing to compensate for the variable error rate performance and added delay of the space link. With regard to the TV example, there are three international analog standards available (NTSC, PAL, and SECAM), and this is being expanded through the new digital TV (DTV) standards, which are all based on MPEG 2 but may not be identical in detail.

2.3.5 Satellite Systems Engineering and Operation

The daunting task of creating an effective satellite communication system involves many disciplines. The integration of the ground and space segments is crucial, as is the proper interfacing of the system with the user environment on the ground. We have specialists who deal with each of these elements and with the individual components of each element. But it is the job of the systems engineer to understand how all of the pieces work together to meet the overall requirements of the project and operation.

We refer here specifically to a satellite systems engineer, someone who understands the functionality of both the space and ground segments. This kind of expertise is not easy to come by, and in fact many who claim to be qualified are not. In many ways, it is the purpose of this (and the previous) book to create a foundation for entering the field of satellite systems engineering. Any ground segment project will, of necessity, require that the satellite systems engineering function be performed properly at every key step. Such performance draws heavily from generic systems engineering, but we cannot lose sight of the fact that we are talking about a satellite communication system. The first step in any systems engineering effort is to understand and define the requirements. In commercial satellite communication, this includes the purpose and strategy of the business that the system supports. Many system engineers work as part of business development because of the close coupling between the two functions during the formative phase. Because of the many tradeoffs that will be required as the design progresses, we must have a complete knowledge of the technology options and their characteristics (technical and financial).

In this author's experience, it is wise to enlist the talent and resources of outstanding analysts who can put the problem down on paper (and on computer) quickly and produce usable results. These studies form the foundation of the "trade space" for the project. By trade space we mean the collection of technology and performance options that we have considered in evaluating how to achieve the system objectives. Then we look at different ways to design the overall systems and their major components (e.g., the earth stations).

Any good systems engineer can do this, at least in principle. However, a satellite systems engineer is one who understands how the space segment is to be included. Through education and practical experience, the satellite systems engineer can investigate the best satellite design (whether GEO or non-GEO), frequency band, and MA method. By performing the link analyses and sizing exercises, he or she can find the optimum arrangement of earth stations in the overall network.

The operation of the ground segment and earth stations is another critical aspect of the overall system, since this determines the quality of service (QoS) and impacts the financial performance of the business as well. Examples of QoS factors include:

- System availability, measure in percent of time that the service is up and operating;

- Data throughput;

- Connection time (for connection-oriented services) and rate (accounting for frequency of busy signals and dropped calls);

- Information transfer delay (also called latency);

- BER as a function of time;

- Qualitative factors such as customer satisfaction rating.

Many of these factors are set by the design of the ground segment and are under the control of the systems and earth station engineers. However, others are the result of how well the ground segment is operated and maintained, which is something that depends on the people who perform these functions during its lifetime. The operation and maintenance (O&M) of ground segment and earth stations is a complex subject, which is treated in more detail in Chapter 11.

References

[1] Elbert, Bruce R., *Introduction to Satellite Communication*, 2d ed., Norwood, MA: Artech House, 1999.

[2] Elbert, Bruce R., *The Satellite Communication Applications Handbook*, Norwood, MA: Artech House, 1997.

[3] Freeman, Roger L., *Telecommunications Transmission Handbook*, 4th ed., New York: Wiley, 1998.

[4] Jasik, Harry, "Fundamentals of Antennas," *Antenna Engineering Handbook*, 3d ed., ed. Richard C. Johnson, New York: McGraw-Hill, 1993, pp. 2–39.

[5] Flock, Warren L., *Propagation Effects on Satellite Systems at Frequencies Below 10 GHz: A Handbook for Satellite Systems Design*, NASA Reference Publication 1108(02), Washington, D.C.: National Aeronautics and Space Administration, 1987, pp. 3–12.

[6] Schmidt, W.G., "The Application of TDMA to the Intelsat IV Satellite Series," *COMSAT Technical Review*, Vol. 3, No. 2, Fall 1973, p. 257.

[7] Suyderhoud, H. G., Jankowski, J. A., and Ridings, R. P., "Results and Analysis of the Speech Predictive Encoding Communications System Field Trial," *COMSAT Technical Review,* Vol. 4, No. 2, Fall 1974, p. 371.

[8] Larson, Lawrence E., ed., *RF and Microwave Circuit Design for Wireless Communications,* Norwood, MA: Artech House, 1997.

[9] Glisic, Savo, and Branka Vucetic, *Spread Spectrum CDMA Systems for Wireless Communications,* Norwood, MA Artech House, 1997.

[10] Balston, D. M., and R. C. V. Macario, eds., *Cellular Radio Systems*, Norwood, MA: Artech House, 1993.

3

Space-Ground Interface Requirements

The ground segments, major earth stations, and user terminals that we address in this book are all designed to provide commercial communications services, the majority of which are normally associated with terrestrial networks. This must be accomplished in a manner that is essentially transparent to the user, meaning that subscribers and customers should not be exposed to the unique technical and operational aspects of the satellites that are used as radio repeaters. Systems and ground segment engineers need to work hard to attain this often difficult objective, because the space segment by its very nature is different from terrestrial means of transmission (e.g., cable, line-of-sight microwave, cellular, and terrestrial broadcasting).

The first step is to understand the requirements from the standpoint of the end-to-end services provided by the overall space/ground system. Our perspective for this discussion is the ground segment—what has been termed the "ground game" [1]. There are many business issues that providers of ground-based satellite communication hardware and services must address, relating to selection of appropriate applications, marketing and sales, sourcing of suppliers, risk management, and integration with the overall telecommunications environment [2]. Readers needing more detail on the applications themselves can find it in our previous work [3].

Readers are no doubt aware that communications satellites operate in three basic orbital arrangements and that each has benefits and operational concerns related to the provision of services to the ground segment. As determined by Kepler's third law, orbit period, and hence time in ground view, is proportional to the mean radius of the orbit raised to power 3/2 (Figure 3.1). The basic properties of the three classes of communications satellites are reviewed below.

- *Low earth orbit* (LEO), ranging between 800 and 1,200 km (corresponding to orbital periods of 1.6 and 1.9 hours, respectively). If the satellites are intended for global coverage (which is the most common service mode for communications services), then these orbits must either be polar or highly inclined to serve users in northern and southern latitudes. As a rule of thumb, the orbit inclination must be approximately equal to the optimum latitude for services. Locations in western Europe and North America may be favored at the expense of equatorial regions such as Asia, Africa, and Central and South America. LEO constellations, as they are called, may be fully populated with satellites (between about 50 and 150) to provide continuous service. A given satellite is able to serve an earth station only for several minutes, a duration further reduced if that earth station must simultaneously connect to another through the same satellite (a difficulty resolved by intersatellite relays of the type found

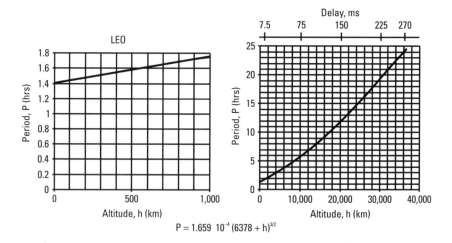

$$P = 1.659 \ 10^{-4} (6378 + h)^{3/2}$$

Figure 3.1 Orbit period and propagation delay for various satellite altitudes, based on Kepler's third law of planetary motion.

in Iridium). Any LEO network requires a system of handover from satellite to satellite, akin to what is done in cellular radio telephone networks to prevent a dropped call when a subscriber moves from one cell to the next.

- *Medium earth orbit* (MEO), at an altitude of greater than about 10,000 km. Orbit period is of the order of six hours and so the satellite will remain in view throughout a majority of connection types (but not for continuous services like DTH and private backbone networks). Also, the altitude, being much higher, allows a given satellite to see a much larger portion of the earth and to thereby serve a greater quantity of users. These factors, taken together, mean that many fewer satellites are required (12 to 24) which may be of greater capacity than LEO spacecraft. The other factor to consider is propagation delay (indicated in Figure 3.1), which is much more significant than in LEO but still reasonably low in relation to the GEO satellites to be discussed next. Both the LEO and MEO configurations permit multiple hops (i.e., relaying through two satellites with a gateway earth station between) because of the relatively brief delay. Such is not the case for GEO satellites, as will be discussed later in the chapter.

- *Geostationary earth orbit* (GEO), a true equatorial orbit with the 24-hour period needed to synchronize with the earth's rotation. At a mean altitude of 35,788.293 km (22,237.814 miles), satellites must still by controlled (stationkept) to correct for gravitational forces that disturb the desired geometry. North-south stationkeeping is used to control the inclination of the orbit (to keep it equatorial) and east-west stationkeeping is used to hold the assigned longitude. A typical stationkeeping tolerance of 0.1° keeps the satellite well within the beamwidth of all but the largest earth station antennas at frequencies up to 20 GHz. Since north-south stationkeeping uses up most of the available fuel, it is a common practice to extend orbit lifetime by ceasing north-south corrections and continuing only with east-west corrections for several more years (until either the remaining fuel is exhausted or a degradation of a critical component renders the spacecraft useless). With inclination greater than about 0.5°, earth station antennas will probably have to track the satellite. The general class of inclined 24-hour orbits, which may be elliptical, is called geosynchronous.

3.1   GEO Satellites and Orbit Slots

We begin with a discussion of GEO, only because this is the foundation of the commercial communication satellite industry. GEO satellite networks need fewer satellites, and ground antennas, while larger than their LEO and MEO counterparts, need not include tracking systems.

3.1.1   Link Characteristics with Link Budgets

GEO satellite links are the easiest to understand and apply because of the simplicity of the geometry and their relatively stable nature. As introduced in Chapter 2, it is an easy matter to compute the free space path loss and thus determine the most dominant factor in the overall design. Still, there are several other considerations in using GEO satellite links that can, at times, become overriding.

The easiest way to understand and appreciate the stability and simplicity of GEO links is to examine a set of link budgets for transmission between a pair of fixed earth stations. Figure 3.2 provides the basic end-to-end block diagram of a point-to-multipoint digital TV transmission using MPEG-2 and the Digital Video Broadcasting (DVB) standard. The satellite repeater in this example is of the bent-pipe variety, which means that it provides the following functions for uplinked signals:

- Reception by the spacecraft antenna;
- Low noise amplification;
- Frequency translation to the downlink frequency range (usually without an inversion, so that the bottom of the uplink band appears at the bottom of the downlink band);
- Filtering down to the transponder bandwidth (30 MHz in this example) by the input multiplexer;
- Gain adjustment using one of the following: (1) fixed gain with or without ground adjustment, (2) automatic gain control (AGC), or (3) hard limiting;
- Power amplification using either a traveling wave tube amplifier (TWTA) or solid state power amplifier (SSPA);
- Channel combining by the output multiplexer;
- Transmission within the footprint area by the spacecraft downlink antenna.

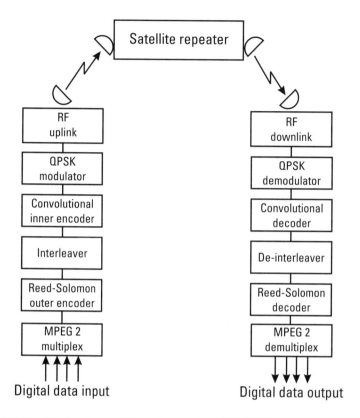

Figure 3.2 Simplified end-to-end block diagram for a DVB MPEG-2 link.

Since in our earlier work [4] we examined each of these elements in considerable detail, we provide only a review here. The satellite repeater does these functions nearly transparently (at least to a first order—more will be said later about how to take the significant higher order effects into account) in the performance of basic functions of ground coverage, power amplification, and frequency translation. The earth stations in the analysis are defined as follows:

- Uplink earth station—a typical BSS broadcast center, similar in concept to that of DIRECTV, discussed in Chapter 1. Each uplinked carrier can support a quantity of digitally compressed TV channels containing full-time video and audio programming. The modulation format in DVB-S (the satellite mode of DVB) is QPSK using concatenated coding (i.e., Reed Solomon outer code followed by a convolutional inner code). A klystron high-power amplifier (HPA) operating at 18 GHz provides RF uplink power to a 13-m transmitting antenna. Tracking and uplink power control (UPC) maintain the received power at the satellite during relative motion of the satellite and during heavy rain along this path, respectively.

- Downlink antenna—a DTH home receiving system with a 45-cm antenna, a true commercial user terminal with many features to simplify installation and use by the subscriber. Within the set-top box are the channel selection, carrier demodulation, error correction, demultiplexing, decryption, MPEG-2 decompression, and conditional access elements needed to reproduce the original programming material.

Basic link budgets for the downlink, uplink, and combined overall link are provided in Tables 3.1, 3.2, and 3.3, respectively. We have included only the top-level performance parameters of the transmit earth station, satellite repeater, and receiving user terminal. These factors provide the basic performance of the end-to-end system; that is, if the three main elements of the system function as assumed, then the service will work accordingly. Tables 3.1 and 3.2 present the stable unfaded condition, which is normal for a GEO link with no rain on the uplink or downlink, for a satellite that is stationkept to remain within the beamwidth of the earth station and user terminal. The links operate in the Region 2 allocation for BSS: downlink between 12.2 and 12.7 GHz and the corresponding uplink between 17.7 and 18.2 GHz. The second half of Table 3.3 accounts for the amount of rain attenuation on the downlink, which would cause a threshold condition, i.e., one in which reception would be just barely acceptable. Rain on the uplink can be countered by uplink power control (this will be discussed further in Chapter 4).

For simplicity, only one channel of communication is considered, and so we have ignored the following effects (these factors will be evaluated in Chapter 4):

- Adjacent channel interference, produced by other carriers on different frequencies received in the same polarization;

- Cross-polarized signals transmitted to and by the same satellite on the same (or adjacent) frequency channels;

- Adjacent satellite interference;

- Linear distortion produced by group delay filters in the earth station and satellite repeater;
- Nonlinear distortion produced by amplifiers, namely the earth station HPA and the satellite TWTA or SSPA, as appropriate.

The downlink (Table 3.1) is straightforward, starting with the satellite repeater HPA output power of 112 watts (produced by a TWTA), and applied to the antenna after passing through output losses of 1.5 dB. The spacecraft antenna gain of 33 dBi corresponds to an area coverage footprint of a relatively large country. The resulting EIRP of 52 dBW is typical of BSS satellites serving nontropical regions, as suggested by the footprint in Figure 3.3.

Free space loss at about 205 dB is the largest single entry; however, it is fixed for the specific combination of earth station and satellite position. The receiving user terminal consists of a nominal 45-cm offset-fed reflector and circularly polarized feed and LNB combination. With about 0.2 dB of loss in

Table 3.1
Link Budget Example for the Downlink at 12.2 GHz (Ku-band BSS)

Link parameter	Value	Units
Transmit power (112 watts)	20.5	dBW
Transmit waveguide losses	1.5	dB
Transmit antenna gain (footprint)	33.0	dBi
Satellite EIRP (toward earth station)	52.0	dBW
Free space loss	205.5	dB
Receive antenna gain (0.45 m)	34.0	dBi
Receive waveguide loss	0.2	dB
Receive carrier power	−118.7	dBW
System noise temperature (150 K)	21.8	dBK
Earth station G/T	12.0	dB/K
Downlink C/T	−141.5	dBW/K
Boltzmann's constant	−228.6	dBW/Hz/K
Bandwidth (30 MHz)	74.8	dB Hz
Noise power	−130.8	dBW
Carrier-to-noise ratio	12.3	dB

Figure 3.3 Sirius 3 BSS Nordic coverage from 5.0° east longitude, indicating EIRP performance at the 52 dBW level (illustration courtesy of NSAB).

the receive feed and a system noise temperature of 150K, the combination produces a receive G/T of 12.0 dB/K. The output of the LNB is a value of C/T of −141.5 dBW/K, obtained with the following simple formula:

$$C/T = EIRP - A + G/T$$

where A is the sum of the path losses (assumed to be free space loss in this example). The corresponding value of C/N ratio is 12.3 dB as measured in the assumed 30 MHz of bandwidth.

The baseband to modulated-carrier design assumed in this example is according to the DVB-S standard for Ku-band links [5]. To establish the carrier bandwidth, we assume an input information rate of 32.4 Mbps, which expands to 46.8 Mbps from Reed-Solomon encoding (188/204) and R = 3/4 FEC encoding, and converted to a carrier bandwidth with a QPSK modulation factor of 1.28/2 = 0.64. We cannot determine if 12.3 dB is adequate for the downlink until the uplink and combined link are considered.

The uplink budget in Table 3.2 follows the same basic format as that described for the downlink. A difference is that the EIRP is produced by the transmitting earth station with 100 watts of HPA power. A considerable

Table 3.2
Link Budget Example for the Uplink at 18.2 GHz (Ku-band BSS)

Link parameter	Value	Units
Transmit power (100 watts)	20.0	dBW
Transmit waveguide losses	3.0	dB
Transmit antenna gain (13m)	65.6	dBi
Earth station EIRP (toward satellite)	82.6	dBW
Free space loss	209.1	dB
Receive antenna gain (footprint)	33.0	dBi
Receive waveguide loss	1.0	dB
Receive carrier power	−94.5	dBW
System noise temperature (450K)	26.5	dBK
Satellite G/T	5.5	dB/K
Uplink C/T	−115.0	dBW/K
Boltzmann's constant	−228.6	dBW/Hz/K
Bandwidth (30 MHz)	74.8	dB Hz
Noise power	−127.3	dBW
Carrier-to-noise ratio	32.8	dB

reserve of uplink power is available from a backed-off klystron-power amplifier (KPA), typically 10 to 15 dB, so that heavy rain on the uplink can be overcome. In addition, the link can obtain greater margin against rain through repeater AGC or limiting in front of the TWTA. The nominal earth station EIRP of approximately 80 dBW, considerably greater than that of the satellite, benefits from the use of a large-diameter antenna (13m), which is continually directed toward the satellite. Satellite G/T, on the other hand, is low (5.5 dB/K) under the assumption that there is a broad coverage footprint to allow access from across the same region as the downlink.

The nominal value of C/N obtain in the uplink is 32.8 dB, considerably greater than the downlink value, causing the uplink to have little effect on overall link and system performance. The reason for this approach is that the uplink is serving a community of literally millions of downlink antennas that are installed and maintained by nontechnical subscribers. Our objective, as service provider, is to assure the highest quality of service using those elements that are under our direct control. The high uplink C/N can be maintained during heavy rain up to the point where our uplink power margin is exhausted (i.e., about 10 dB of attenuation). From that point, the repeater AGC takes over by maintaining constant drive to the TWTA as higher rain attenuation reduces the input power. The uplink C/N will now decline with this loss of input, thus degrading the overall performance. This is another reason that a high value of clear weather uplink C/N is desired. Figure 3.4, which is the cumulative distribution of attenuation at 20.2 GHz at Clarksburg, Maryland, provides some measure of the amount of uplink margin needed for an 18-GHz uplink. For example, one might expect that for this location, about 14 dB of uplink margin will deliver 99.9% availability in most years.

The combined performance of this BSS link example is presented in Table 3.3, which assumes a bent-pipe type of repeater (i.e., one that allows uplink noise to pass through to the downlink). After converting C/N values to true ratios (i.e., $10^{(C/N)/10}$), we obtain the combined C/N as follows:

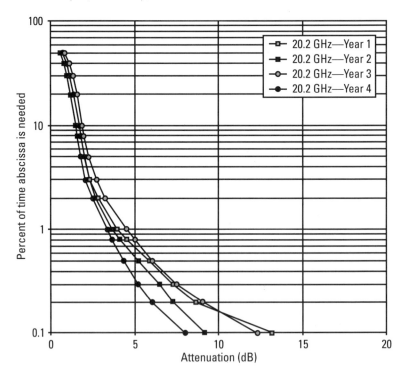

Figure 3.4 Cumulative distribution of attenuation at 20.2 GHz as measured at Clarksburg, Maryland, over a four-year period (illustration courtesy of ACTS Propagation Workshop).

Table 3.3
Combined Downlink and Uplink Example, Ku-band BSS, Clear Air and
Downlink Rain Conditions

Link Parameter	Value	Units
No Fading (clear air)		
Uplink C/N	32.8	dB
Downlink C/N	12.3	dB
Overall link C/N (thermal)	12.3	dB
Carrier-to-interference (C/I)	16.0	dB
Total link C/(N + I)	10.7	dB
Minimum requirement	6.8	dB
Overall system margin	3.9	dB
Faded (downlink rain)		
Uplink C/N	32.8	dB
Downlink C/N (with 3.0 dB fade)	9.3	dB
Overall C/N (thermal)	9.3	dB
Carrier-to-interference (1.3-dB decrease in desired carrier)	14.7	dB
Total link C/(N + I)	8.2	dB
Minimum requirement	6.8	dB
Overall system margin	1.4	dB

$$C/N_{th} = [N_u/C + N_d/C]^{-1}$$

where C/N_{th} is the combined C/N for the thermal noise (i.e., the receiver noise produced in the user terminal and satellite front end), N_u/C is the inverse of the true ratio of the uplink C/N (Table 3.2), and N_d/C is the inverse of the true ratio of the downlink C/N (Table 3.1).

This formula can be extended to account for interference entries from adjacent channels (ACI), cross-polarization (XPOL), adjacent satellites (ASI) and terrestrial interference (TI) sources (an example follows):

$$C/N_{tot} = [N_{th}/C + N_{aci}/C + N_{xpol}/C + N_{asi}/C + N_{ti}/C + \dots]^{-1}$$

Of course, we must convert back to dB as a final step in this calculation.

For the BSS link in question, Table 3.3 evaluates first the combined (thermal) C/N for the unfaded (clear air) condition, which will persist for the majority of the time in all but the rainiest climates in the world.

We have assumed C/I = 16 dB as an allocation for all RF interference. The clear air C/(N + I) of 10.7 dB is then compared to the minimum requirement for our user terminal receiver, assumed to be a QPSK demodulator with R = 3/4 forward error correction as used in the DVB standard (reviewed later in this chapter). For this type of device, the standard allows 6.8 dB for satisfactory operation; less than this value can produce reception difficulties such as dropouts and loss of sync. Our overall system margin is 3.9 dB to take account of rain attenuation on the downlink and other effects (e.g., additional interference, antenna mispointing, and the like).

A condition with 3 dB reduction of C/N due to rain fade is shown at the bottom of Table 3.3. Due to the fact the rain produces both attenuation and an increase in downlink noise due to absorption, the actual rain loss is about 1.3 dB. Adjacent satellite interference is assumed to follow a different path and not be attenuated; therefore, C/N_{asi} is reduced by the amount of rain attenuation (e.g., 1.3 dB). The remaining 1.7 dB of fade is that produced by the increase in system noise temperature, calculated according to:

$$\Delta T = (l - 1 / l)270$$

where l is the rain absorption as a ratio greater than 1 (e.g., 1.35 in this example). The downlink noise power increases by a factor equal to [T + ΔT]/T, which is [150 + 70]/150 = 1.47, or 1.7 dB.

We see that the C/N_{tot} has decreased to 8.2, resulting in a system margin of 1.4 dB. This particular amount of fade is typical of what one would expect in a temperate climate such as western Europe (the last entry in Table 3.3) or the northeastern United States. Having an extra 1.4 dB of margin provides that much more confidence in a system that will adequately serve the customer base.

From the uplink and downlink budgets, we can identify the key characteristics of the transmitting earth station and receiving user terminal (Table 3.4). This example should demonstrate the simplicity with which GEO links can be analyzed and understood. The ground segment design is quite stable, because for most applications antenna tracking is not required. From these tables, we can determine the primary performance requirements for the earth stations and user terminals.

Table 3.4
Primary RF Performance Requirements for the Earth Station and User Terminal
Used in the Link Budget Example

Uplink earth station	Value	Units
Uplink EIRP, single carrier, clear-air operation	82.6	dBW
HPA output power, single carrier, clear-air operation	100	Watts
Uplink power control range	10	dB
HPA size (minimum)	1,000	Watts
Transmit antenna gain	65.6	dBi
Antenna diameter, nominal	13	meters
Maximum EIRP, heavy rain	92.6	dBW
Waveguide loss, maximum	3	dB
User terminal	**Value**	**Units**
Receive G/T	12.0	dB/K
Antenna diameter	45	cm
Antenna temperature	50	K
Receiver noise temperature	90	K
Feed loss noise	10	K
System noise temperature	150	K
Antenna gain, minimum	34.0	dBi
Received C/N at threshold	8.5	dB

3.1.2 Orbit Spacing

Since the vast majority of operating GEO satellites are north-south and east-west stationkept, we need only consider two factors to describe their location and operation: the assigned longitude and the spacing between satellites on the same frequency. The fact that many satellites are visible from a particular ground location provides motivation for repositioning the antenna from time to time. If this is contemplated, then the antenna mount should have the mechanical adjustment range appropriate to the need.

Locating a GEO satellite from the ground is quite straightforward, but can be a daunting task for larger antennas. Figure 3.5 provides a simple nomogram to estimate the earth station elevation and azimuth angles relative to true north. The y-axis indicates the earth station latitude, either north or south, in degrees. The x-axis, on the other hand, called the relative longitude, is the absolute magnitude of the difference in degrees between the earth station longitude and the longitude of the satellite. Thus, if the earth station is at 110 degrees WL and the satellite is at 85 degrees WL, then the relative longitude is 25 degrees (i.e., 110 − 85). The elevation angle, which also determines the slant range, is found from identifying the closest arc (0 degrees is the outermost bounding curve, e.g., pointing the antenna at the horizon). The azimuth is obtained from the radial curves in the figure, taking account of which quadrant the earth station is in as seen from the satellite (see upper right-hand corner of the figure for the key). To continue the example, assume that the earth station is at 40° NL, then our reference is quadrant II in Figure 3.5, since the ground location is to the northwest of the subsatellite point. The corresponding azimuth is 360° − 217° = 143°, and the elevation angle is 38°.

A more effective way to determine where to point a real antenna is to use PC software such as reviewed in section 8.2.2. In addition, many satellite operators provide a pointing calculator on their Web page (for example, see www.panamsat.com). Any of these calculations give nominal pointing to the satellite when it is in its "center of box," meaning the ideal position at the assigned longitude and zero degrees inclination with respect to equatorial orbit. Actual stationkeeping produces error in this position that should be within ±0.1° of the prescribed value. With four or more satellites sharing the same orbit position in some cases, individual nominal positions could be either to the left or right of center. Another strategy developed for large groupings uses variations in inclination and eccentricity, which is more robust. The particular arrangement depends on the operating approach of the operator or the manner in which the slots were assigned by the associated government regulatory body.

Transmitting earth stations and user terminals requires special consideration when employing frequencies at Ku-band and higher. As stated for the digital link budget, uplinks may require UPC to maintain adequate signal level at the satellite and acceptable link availability. Techniques on the satellite that can remove some of this variation include automatic gain control (AGC), limiting, and demodulation/remodulation in a digital processor. This is a complex question requiring careful analysis of the system and its elements. For links at Ka-band and higher, other techniques may be required if the resulting availability is not acceptable. An example of the performance of a given Ka-band path measured in the Washington, D.C., area by COMSAT

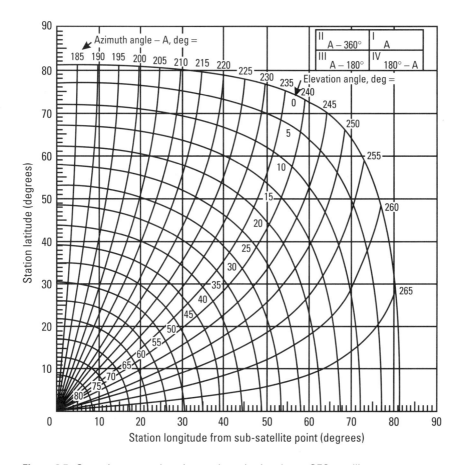

Figure 3.5 Ground antenna elevation angle and azimuth to a GEO satellite.

Laboratories is provided in Figure 3.4. One technique to overcome severe uplink rain-fade is to employ ground diversity, where two or more sites are sufficiently separate so that the incidence of both sites experiencing a deep fade is reduced greatly. This can be of the order of 20 km or more, and so it imposes both investment and operational constraints on the system.

A last factor is the requirement of satellite spacing itself, which is determined by the beamwidth and sidelobe performance of the transmitting and receiving earth stations associated with each satellite. Adequate carrier-to-interference ratio (C/I) can be obtained at the expense of increasing earth station antenna diameter. A critical concern is uplink interference, which is presented in Figure 3.6 for C-band earth station antennas of 5-m and 10-m diameter (orbit spacing and power level difference are the variables). The

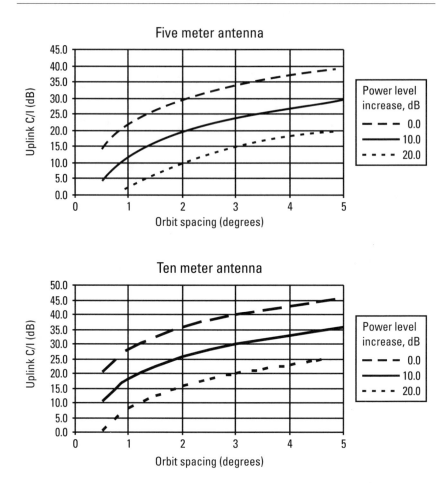

Figure 3.6 Uplink C/I due to a C-band transmitting earth station for either 5-m or 10-m reflector diameter, as a function of orbit spacing and uplink power level increase (assumes 10% increase in offset angle due to ground geometry).

governing relationships are: (1) the sidelobe radiation envelope, which was introduced in Chapter 1, and (2) the gain of the desired and interfering satellites as determined by their respective footprint coverages.

3.2 Non-GEO Constellations

Non-GEO satellites have been applied to satellite communication for many decades. For example, the Molniya satellites operated by Russia during the

Soviet era proved that one could derive practical services from satellites in elongated elliptical orbits. A satellite could be used only during the period when it was far from the earth, giving path lengths as long as or longer than for GEO. Also, the 12-hour period of the orbit requires that two satellites be utilized for continuous service (earth stations must hand off between satellites, using dual antennas). For the country with the largest geographical area, and one substantially at higher latitudes, Molniya offered benefits in terms of high elevation angles and extended periods when one satellite could be viewed from both ends of the country. But this particular system outlived its usefulness in Russia's more recent history.

The Iridium constellation of 66 LEO satellites established non-GEO arrangements as technically viable for mobile communications. Readers are familiar with the approach taken by Motorola in the development of this project, but the system entered bankruptcy and its future in was in question at the time of this writing. But what we are interested in here is the way that such constellations affect the manner that the ground segment is designed to serve users.

3.2.1 Constellation Characteristics

The basic relationship governing the application of non-GEO satellites is provided in Table 3.5, which indicates the altitude, required number of satellites for continuous service, orbit period, maximum time in view of a subscriber, and mean propagation delay. We have selected typical configurations for discussion, although other variations are certainly possible. For example, Globalstar is an important LEO constellation that uses 48 satellites at 1,400-km altitude in orbits that are inclined 52° to improve coverage of the

Table 3.5
Key Parameters for Various Orbit Constellations

Orbit	Altitude (km)	Period (hrs)	Number of Satellites	Nominal Inclination	Maximum Time in View (hrs)	Single Hop Propagation Delay (ms)
GEO	35,788	24.0	3	0°	24	260
Molniya	40,000 (apogee)	12.0	2	65°	6.0	300
MEO	10,355	5.8	12	45°	2.0	40
LEO	780	1.4	66	86.4°	0.2	5

inhabited latitudes of the planet. Globalstar does not cover the poles as well as Iridium, something that GEO satellites cannot accomplish at all. This table considers physical aspects rather than characteristics of the link and service itself. For example, a LEO constellation offers global coverage (if the orbits have sufficiently high inclination), but whether the user terminal antennas need to track or not depends on the frequency band and bandwidth demands of the link. An L-band system supporting narrowband communication, as in Iridium, can get by with a near-omnidirectional antenna on the handheld terminal. As the bandwidth and frequency are increased, the link demands significant antenna gain, which can only be provided using a directional beam of some kind. Once the beam is directional, it must point toward the operating satellite.

The maximum size of the footprint is determined by the radius of the circle of potential coverage centered on the subsatellite point (i.e., the point on the ground directly below the instantaneous position of the satellite). The Molniya satellite, at approximately 40,000 km maximum altitude, has a larger footprint than GEO, while LEO satellites are limited to coverage radius of the order of 1,000 km. This area may be furthur subdivided into cells. The coverage areas must overlap and move in an orchestrated fashion, based on the orbit arrangement. Progression of the satellites in orbit can be shown using computer simulation and visualization with a software tool such as Satellite Toolkit from Analytical Graphics (www.stk.com), discussed in Section 8.2.1.

3.2.2 Continuity-of-Service Issues

Maintaining service continuity is a critical issue in non-GEO constellations because superior performance requires more satellites and gateways, and hence greater investment. The most critical requirement is sufficiently high elevation angle from the ground to clear local terrain obstacles, which dictates the minimum number of satellites to achieve 100% continuity of service. This can be appreciated by reviewing two different architectures: the bent-pipe/ground hub approach of Globalstar and the broadband routing/intersatellite relay approach contemplated by Teledesic. The bent-pipe/ground hub approach connects user terminal traffic to the satellite and a gateway earth station. Information can then pass from the user terminal over the satellite, through the gateway and onto the network control center (NCC) via another network (presumably terrestrial). The quality of the user-to-gateway connection depends on the stability of the link and the ability of the overall network to assure that this connection won't be broken during handover to the next in-view satellite. For satellites of the simple bent-pipe design, handover is accomplished within the gateway earth stations that perform all of the signal processing and call management functions. The technique involves the gateway receiving the same signal from two satellites and selecting the strongest of the two at every instant in time. Tracking of the satellites by gateway antennas is, of course, a necessity. User terminal antennas of the type used in Globalstar do not require tracking and handover, and so their design is much like that of a conventional digital cellphone (except for the required upward view, as discussed in Chapter 9).

In contrast, the tracking and handover requirements of high data rate Ka-band user terminals place considerable demands on the design, since these devices must be affordable to many millions of business and individual subscribers. While the last comment relates to non-GEO systems, the stationkeeping properties of a GEO satellite can eliminate tracking requirements for user terminals.

Gateways may be installed to provide access to fixed users, such as through the Internet, PSTN, or private networks of some type. The network is robust in that, with an excess of satellites and links between them, a communication path can be established under nearly any circumstance. The one exception is when the user terminal is blocked from view of any satellite. This sort of thing is a normal characteristic of communication using a non-GEO constellation, but can possibly be made acceptable by supplying an excess of satellites and link margin.

3.2.3 Link Characteristics for Non-GEO Systems

Service to the ground segment of a non-GEO system depends in large part on the nature of the link, which is shorter but decidedly more dynamic than that of its GEO counterpart. Current LEO MSS systems have the advantage that user terminals do not need to track the satellites as they pass and as the link connection is transferred during calls. Fading on these links is primarily the result of multipath propagation from reflections off of buildings and the ground, and local terrain blockage. As we move to broadband services that require user terminals with tracking beams, the situation becomes much more difficult to address. The use of Ku and Ka bands adds the factor of rain attenuation.

Due to the different orbital altitudes, there is a "near-far" problem similar to that experienced in terrestrial cellular telephone (e.g., the path length for a user close to a base station as compared with one at the edge of the cell). The actual dB variation is plotted in Figure 3.7 for GEO and

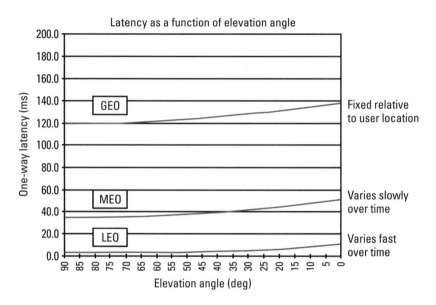

Figure 3.7 A GEO system presents considerably more range distance and thus path loss than a LEO, but the LEO must support a wider dynamic range (*Source:* Jeffrey Maul, Arthur D. Little, Inc.).

typical MEO and LEO altitudes. The corresponding near-far dB ranges are 1.3 dB, 3.5 dB, and 12.3 dB, respectively. With a 12.3-dB range of variation across its coverage, the LEO satellite provides a decidedly different kind of service than one is accustomed to from GEO satellites with their essentially constant service level across the footprint (neglecting the effect of the gain variation of the satellite antenna pattern). In practice, the near-far variation can be countered by increasing satellite antenna gain as one looks from center to edge of coverage, but this can be difficult to do on a technical basis since antennas naturally tend to exhibit defocusing loss as the beam is scanned off the main axis.

As a final consideration for the non-GEO type of link, we indicated in Table 3.5 the time that a satellite is in view from a particular user on the ground. This time is further reduced if an associated gateway or control station must be in view as well. The LEO satellite will appear to spend more of its time at the lower elevation angles than at the higher. This, of course, is less desirable because of the greater potential for terrain blockage and rain attenuation for Ku and Ka band links.

This completes our brief review of requirements due to the basic physical characteristics of space segments (GEO, MEO, and LEO). At this point, we transition to a discussion of the typical types of services that satellite communication systems currently are able to provide. These are divided into two broad categories based on fundamental connectivity: two-way communications services (Chapter 4) and one-way broadcasting (Chapter 5). As a result, the applications they support are very different in terms of capability and business model.

References

[1]   Watts, T. W., and D. A. Freeman, "Satellite Communications—Instant Infrastructure," Bear-Stearns Equity Research, November 5, 1996.

[2]   Frost and Sullivan, "Worldwide Satellite Earth Station Component Markets," Report No. 5428-60, 1997.

[3]   Elbert, Bruce R., *The Satellite Communication Applications Handbook*, Norwood, MA: Artech House, 1997.

[4]   Elbert, Bruce R., *Introduction to Satellite Communication*, 2d ed., Norwood, MA: Artech House, 1999.

[5]   European Broadcasting Union, European Telecommunications Standards Institute, "Digital Video Broadcasting (DVB); Framing Structure, Channel Coding and Modulation for 11/12 GHz Satellite Services," DN 300421, V1.1.2 (1997–98), p 20.

4

Two-Way Communications Service Requirements

The foundation of most telecommunication services is the two-way interactive mode, typified by a telephone conversation between two people. This is, after all, something that we learn from birth, and provides the principle means of exchanging information, views, ideas, and instructions during the normal course of our lives. This section addresses four implementations of satellite communication services: voice networks between fixed points on the ground, narrowband data communications using very small aperture terminals (VSATs), and mobile satellite communications to accomplish in remote regions what cellular and personal communications services/personal communications networks (PCS/PCN) can in developed areas, and broadband services to support or supplement the Internet. Table 4.1 provides a summary of the capabilities of these services as they exist at the time of this writing, while the discussion that follows reviews their characteristics as they relate to the ground segment.

The three types of multiple access (MA) that can be applied to these networks include frequency division multiple access (FDMA), time division multiple access (TDMA), and code division multiple access (CDMA), reviewed in Chapter 2. As reviewed in Table 4.2, each has its merits and issues as far as application in satellite networks. FDMA is appropriate for sharing of spectrum and power within a transponder; it may be used in

Table 4.1
Characteristics of Two-Way Interactive Services Provided by the Ground Segment

Characteristic	Orbit Configuration	Bandwidth per Channel	Multiple Access	Application	Ground Segment Implications
Fixed voice networks	GEO	8 to 64 Kbps	FDMA and TDMA	International telephone trunks to extend the global network, and thin route and rural services in developing regions.	May employ conventional bent-pipe satellites as well as some processing satellites that can reduce earth station size and cost.
VSAT data networks	GEO	64 to 512 Kbps	TDMA and CDMA	Private data networks within countries and regional networks to promote businesses that require greater extent.	Employs partial transponder capacity. Users share resources of the hub and network control.
Mobile satellite communication	GEO, MEO, and LEO	4 to 20 Kbps	FDMA, TDMA, and CDMA	National, regional, and international roaming for mobile telephone and low-speed data, augmenting the global fixed and mobile networks.	Requires a total integrated system with a common air interface. Major operators develop, manufacture, and distribute user terminals.
Interactive broadband	GEO, MEO, and LEO	64 Kbps to 155 Mbps	FDMA and TDMA	Public Internet and multimedia applications, local and backbone.	Requires powerful uplink or high-gain satellite antenna; service to fixed terminals, hubs, and mobile terminals on vehicles, aircraft, and ships.

Table 4.2
Characteristics and Application of Multiple Access Methods in Satellite Communications
Ground Segments

Multiple Access Method	Principle Application	Benefits	Issues
FDMA	Sharing of transponder bandwidth and power	Minimum coordination among users	Power sharing, adjacent channel interference, and intermodulation
TDMA	Digital multiple access networks with consistent timing and coordination	Efficient in terms of bandwidth and power usage; high degree of integration with networks	Limited ability to control interference using separate timing or frequencies; requires timing reference
CDMA	High RF interference environments; variable bandwidth and power availability	Can tolerate self and external interference; little coordination among users; low potential to cause interference	Power control; adequate bandwidth for spreading

SCPC DAMA networks and for operating different applications within the same bandwidth. TDMA, on the other hand, requires tight coordination of timing and bandwidth usage, and may be susceptible to interference (as is FDMA). It is beneficial for digital networks because it allows multiple logical channels within the time frame and integrates well with control channels and back-end systems. The third method, CDMA, has been touted for its efficiency in environments characterized by heavy interference from internal and external sources. One benefit is that the transmissions can be adapted to the instantaneous environment, which is often required in a real-world network.

Because of the highly variable nature of the different applications and environments, these methods represent the principle points of departure for the design of appropriate network architectures and rely upon the design and manufacturing capability of industry specialists such as Motorola, Hughes Network Systems, Gilat, Nokia, Ericsson, Alcatel, and NEC. In our experience, the selection of the optimum MA technique relies more on the experience base and economics of the particular supplier and operator as opposed to the ideal technical features that one might analyze on a sheet of paper.

Any architecture that uses a start topology (or those meshes that require connections back to the PSTN or the Internet) will require one or more hub earth stations. This is illustrated by a typical GEO MSS system, shown in Figure 4.1, that has three types of earth stations: the L-band user terminal, the C-band gateway earth station (facilities in three different countries or regions are indicated), and the shared Network Management Center and Subscriber Management System. This particular network does not allow direct UT-to-UT connections as it only offers services between users and the PSTN via the gateways. We see that this architecture offers tight control (vital functions, not to be underestimated) and management of the network, including measurement of usage and billing.

The following sections review some of the detailed requirements for these types of networks that can be served through a satellite communication ground segment. This is not meant as a comprehensive treatment but rather an introduction to this type of investigation.

4.1   Fixed Telephony Voice Networks

Voice services are inherently interactive in nature and so the service must adhere to user expectations and quality objectives. We have as a basis the work done over the past decades at major institutions like Bell Laboratories in the United States and European industrial and intergovernmental

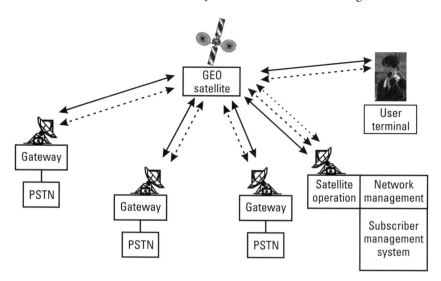

Figure 4.1 GEO mobile satellite and network architecture.

organizations that seek to standardize products and services and promote international business and social development. Reviewed below is a body of engineering practice that has developed from a terrestrial telecommunications foundation and has been adapted to the unique aspects of satellite communications ground segments [1].

At the same time that voice networks become more readily available and lower in cost to use, several important trends have appeared which are causing a paradigm shift in the engineering and operation of the global telephone infrastructure. As summarized in Table 4.3, these result in significant enhancement to the quality, affordability, and versatility of voice services from terrestrial networks. One of the more interesting innovations is

Table 4.3
Trends in the Use of and Requirements for Public Telephone Networks

User Requirement	Approach	Impact
Voice telephony	Fundamental to the design and operation.	None (intrinsic to the current design and operation).
Fax transmission	Users employ Group 3 fax machines, with internal 9.6 to 14.4 Kbps modems.	Basically the same as voice telephony.
Advanced calling features	Digital telephone switches provide a range of features, like call waiting, voicemail, conferencing, blocking, caller ID, etc.	Available in developed countries when new telephone switches are introduced. Generally not available in developing regions trying to meet the basic needs of new subscribers.
Intelligent network features	Signaling Systems No. 7 (SS-7) supports online credit card calling, virtual private networks, integration of domestic with international networks, and improved network management.	Facilitates smooth interoperation of global telephone network; improves reliability of networks; allows greater usage by subscribers and reduced telephone service cost.
Internet access	E-mail and Web browsing extends connect times over access lines and local trunks; introduction of Voiceover IP (IPVoIP).	Local telephone service shifting to different models of service to allow extended connect times and greater bandwidth.
Impact of wireless networks	More people depending on wireless telephony for personal communications.	Reduced local calling for voice with possible implications for future network design.

Voiceover IP (VoIP), an evolving standard for adding telephone services to the Internet space. Satellite voice networks can still offer a relatively low-cost alternative for rural and developing regions until state-of-the-art terrestrial networks can be justified by the local economy.

The engineering and economic principles that underlie the design and operation of public telephone facilities have evolved over the 100 years that these services have been available. These are summarized in the following paragraphs as related to satellite networks that offer an alternative to terrestrial networks (or a complement, where appropriate). As the previous discussion indicates, the rules for network design are constantly being rewritten due to technology and economic/market factors. For example, in the United States the cost of long-distance service has dropped rapidly and subscribers can enjoy flat rates for nationwide calling of $.07 per minute or less, any time and to anywhere. Greater adoption of VoIP promises changing rates of this order on a global basis. At this level, subscribers need not even hesitate to make a call, day or night.

The public switched telephone network (PSTN) represents the common denominator for telecommunications throughout the world. To understand how satellite telephony services can interface with typical subscribers, we have to examine the architecture and interface arrangements of the PSTN. The basic interface arrangement for telephony may be split into the subscriber local loop and the trunks that connect between telephone-switching exchanges. Typical local loop configurations include two-wire analog local loop, four-wire analog loop to a private branch exchange (PBX), the digital loop utilizing the Integrated Services Digital Networks (ISDN) interface at 144 kbps, and digital loop carrier using time division multiplex (TDM) that provides either 24 (T1) or 30 (E1) digital voice channels.

Within the central office of the local telephone company we find the opposite end of the local loop connection, wherein a common digital format is used. Figure 4.2 provides a block diagram of a typical digital switching office used at the local level. This particular arrangement by Lucent Technologies is adaptable to all three general classes of exchanges: PBX, local exchange, and toll (long distance). Digital circuit switching operates on voice channels at 64 kbps, and offers digital trunks to interconnect with other exchanges. The majority of new installations in developed and developing countries employ digital fiber-optic transmission for these trunks, which assures high capacity and high quality for all interconnected services. In time, conventional time division switching with fixed bandwidth per channel will give way to dynamic packet routing using asynchronous transfer mode (ATM), but the timetable for this remains uncertain at the time of this writing.

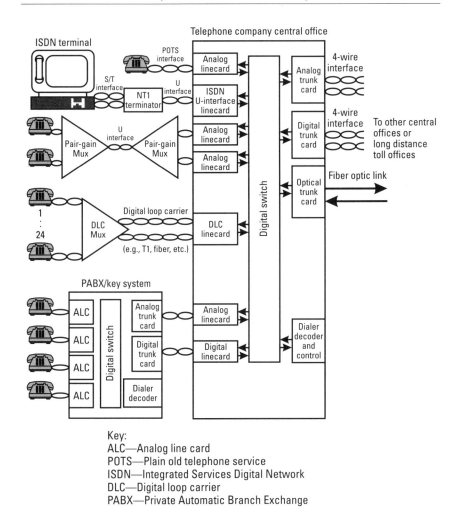

Figure 4.2 Interface arrangements for the PSTN. (Reference [2], with permission.)

An important feature to be provided along with the voice transmission is common channel signaling using Signaling System No. 7. This is a high-speed, complex data communication network architecture that supports the widest range of calling services and network management.

4.1.1 Bandwidth and Quality

Telephone instruments on the PSTN are predominantly analog two-wire devices. As shown in Figure 4.3, the network interface converts from two

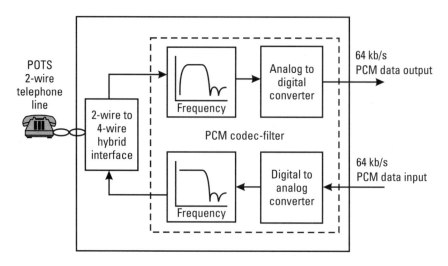

Figure 4.3 An analog telephone line-card that interfaces with the local loop on a two-wire basis and provides 64 Kbps PCM data input/output. (Reference [2], with permission.)

wires to four wires using a balanced transformer called a hybrid. As we will see later, the hybrid is the device most likely to produce echo impairment in long-distance telephone calls. The sending end of the circuit (incoming to the exchange) is filtered to restrict the channel bandwidth to the range 300 to 3,400 Hz. This allows the analog-to-digital (A/D) converter to sample at 8,000 samples per second, which is slightly more than twice the highest voice frequency. Using standard eight-bit-per-sample pulse code modulation (PCM), the output data rate is 64 kbps. (A technique called companding enhances the perceived signal-to-noise ratio, compensating for the imprecision inherent in A/D conversion.) On the receive side, the incoming 64 kbps bit stream is converted to analog and filtered to pass only the frequencies below 3,400 Hz. Receive analog information is inserted back into the two-wire line using the same hybrid.

Voice band channels are useful for relatively low data rate applications like fax, E-mail, and low-speed Internet access (i.e., less than 64 kbps). What makes this possible is the passband response and amplitude linearity of the PSTN, indicated by the attenuation mask in Figure 4.4. The dynamic properties of the channel for variable power levels, shown in Figure 4.5, allow analog telephone circuits to be very adaptable (which is an important reason they continue to be used extensively). The dynamic range is improved through companding, where Mu-law is used in North America and A-law in most other countries (particularly Europe). These characteristics are

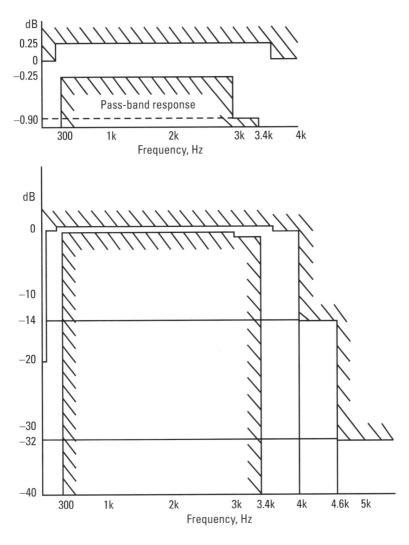

Figure 4.4 Typical frequency response for the input filter within the telephone line-card: passband and stopband. (Same attribute as Figure 4.3) Reference [2], with permission.

important to providing convenient data communication connections over the dial-up PSTN, owing in particular to the popularity of the V.90 standard for 56 Kbps modems.

Another important trend in the use of the analog local loop is the addition of high-speed access to the Internet through digital subscriber line/ loop (DSL) technology. The family of formats and solutions indicated in

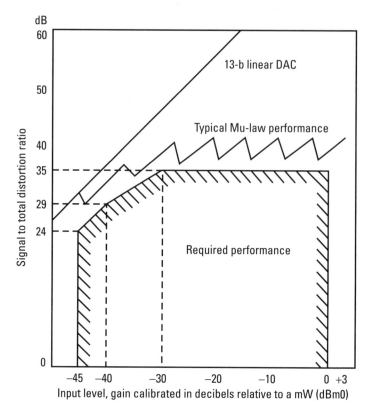

Figure 4.5 Signal-to-quantization distortion performance for Mu-law companded and 13-bit linear digital-to-analog conversion as compared with telephone network requirements. (Same attribute as Figure 4.3) Reference [2], with permission.

Table 4.4 defines most of the implementations that allow data rates in the hundreds to millions of bits per second to pass in both directions between the home and telephone central office typically using the existing copper twisted pair wire [3]. These break down into two categories: asymmetric DSL (ADSL), which provides a higher data rate for downstream (central office to home) than for upstream; and symmetric DSL (SDSL), which assures the same speed in both directions. Most implementations are going to ADSL to be consistent with the Web surfing tendencies of home users. Business users, on the other hand, may require exactly the opposite balance and hence SDSL or dedicated T1/E1 may be preferred. Since most forms of DSL share the bandwidth on a twisted pair of the voice channel, there is a distance restriction on the loop length. While these do not interface directly

Table 4.4

Implementations of DSL

Flavors	Properties	Reference
Asymmetric		
ADSL (full rate)	Offers differing upstream and downstream speed (up to 7 Mbps); allows high speed and voice to share the same line (analog and ISDN); requires special modem	ITU-T Recommendation G.992.1; ANSI Standard T1.413-1998
G.Lite ADSL	Medium bandwidth version for consumer market at speeds up to 1.5 Mbps downstream and 500 Kbps upstream	ITU-T Recommendation G922.2
Rate adaptive DSL (RADSL)	Nonstandard ADSL (standard ADSL also is adaptive)	
Very-high-rate DSL (VDSL)	Up to 26 Mbps on short local loops, within telephone plant; could be installed for VOD	
Symmetric		
Symmetric DSL (SDSL)	Being installed in U.S. by competitive local exchange carriers (CLECs); offers service comparable to LANs; popular in Europe	ETSI standard in preparation
High-data-rate DSL (HDSL)	Originates from 1980s; offers 1.5 or 2.3 Mbps without standard telephone service; replacement for T1/E1; uses two twisted pairs	ETSI and ITU standards in development
Second-generation HDSL (HDSL-2)	Replacement for T1 service (1.5 Mbps) without standard telephone service; uses one twisted pair	ANSI standard
Integrated services digital network DSL (IDSL)	Up to 144 Kbps using existing phone lines, through digital loop carrier, but provides "always-on" ISP service	

with telephone satellite networks, they represent data requirements that may need to be accommodated in some manner.

The digital standard for PSTN access at the subscriber level is ISDN, a service that provides 144 Kbps of active data subdivided into two 64-Kbps circuit-switched "bearer" channels plus one "data" channel (i.e., 2B + D). The standard provides for several interfaces, the selected one depending on the application and level within the network. Illustrated in Figure 4.6 is the reference model for the user interface, which provides for the following [4]:

- Terminal equipment type 1 (TE1), the user system consisting of an ISDN telephone instrument (which supports 144 kbps data and the 2B + D service), compliant fax machines and computer devices capable of using the ISDN service to establish 64 kbps connections via the 16 kbps D signaling channel.

- Terminal equipment type 2 (TE2), which are conventional consumer devices that cannot interface directly on a digital basis, including analog telephones, modems (e.g., V.90), and data terminal equipment (DTE) for other standards, notably X.25.

- Terminal adaptor (TA), an adaptation device to allow connection of non-ISDN devices that do not fall under the TE2 set of definitions. This is a catchall that requires a custom-tailored interface to perform such functions as A/D conversion, protocol adaptation, and channel signaling. An example is an ISDN custom PCI card that plugs into the PC. For a large enough user community, the particular TA device could become a de facto standard.

- Network termination 1 (NT1), the basic interface between the user terminal and the line to the central office. It supports the particular type of physical layer transmission medium being employed between the user and operator.

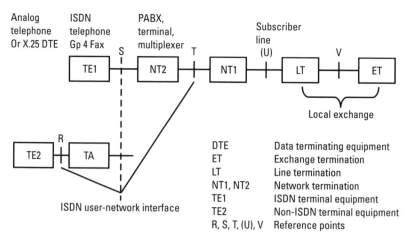

Figure 4.6 Standard user interface definitions for narrow-band ISDN.

- Line termination (LT), the other end of the link from the NT1, providing the interface to the exchange.

- Exchange termination (ET), the terrestrial interface to the ISDN-compatible telephone exchange, providing digital transmission at the bearer rate of 64 kbps and signaling via the packet switched network (using SS-7 within the PSTN).

These interfaces are external to the ISDN switching fabric and digital trunks within the network itself. The ISDN standard was to have become pervasive in the digitized PSTN, but many telephone operators have been reluctant to include the necessary equipment in the telephone exchanges and local loop facilities. The advantage of 2B + D lies mainly with its ability to provide switched 64 kbps of clear-channel data. On the other hand, the wide availability of inexpensive V.90 modems and development of DSL have diminished the appeal of ISDN as it is currently defined. However, the infrastructure aspects of ISDN have proven successful around the world, particularly from a switching and network management standpoint.

The 64-kbps voice channel discussed previously delivers the highest-quality service for telephony. With the adoption of digital transmission from end to end, subscribers have become accustomed to clear-channel service without noise and with a minimum of disruption. Prior to digital fiber-optic and satellite services, networks were analog in nature and suffered from a variety of impairments. This is largely behind us; however, impaired voice communication has returned in the form of the digitally compressed speech used in mobile networks and via the Internet through VoIP.

The ITU has provided a process for evaluating the quality of voice channels that are impaired as compared to what subscribers expect. This is the mean opinion score (MOS) technique, embodied in the P series of recommendations [5]. A group of test users try out the subject telephone link, either on a listen-only basis or in an interactive two-way conversation. After completing the session, they rate the quality according to the following scale:

5. Excellent
4. Good
3. Fair
2. Poor
1. Bad

The scores are then averaged to produce the MOS for the particular run of the overall experiment. This rating system has proven quite reliable over the years and has maintained its popularity in the satellite communication and wireless industries. The MOS of a standard 64-kbps PCM telephone channel generally is slightly higher than 4.0, while significant impairment that reduces the naturalness or consistency of the channel will cause the rating to drop by one or more points. Ratings below 3.0 are generally not indicative of commercial quality. A typical trend curve is provided in Figure 4.7 [6]. Aside from bandwidth and naturalness, one of the most important impairments to a telephone conversation is delay, discussed in the next section.

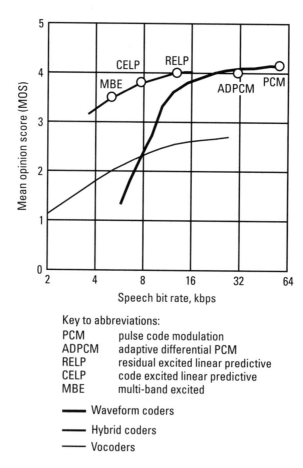

Figure 4.7 Comparison of bit rate and MOS for various digital speech compression techniques.

4.1.2 Delay and Latency

The transmission of communication signals, whether via fiber-optic cable, copper, or radio waves, entails propagation delay. GEO satellites have long been used for telephone services, but it is often pointed out that the 260 ms of delay associated with one hop is a concern in this age of ubiquitous wireless networks and the Internet. One common principle for Iridium and Globalstar is to greatly reduce the end-to-end delay of a simple connection. In this section, we will consider how delay impacts voice services and what facilities are available to make satellite communication links suitable for commercial telephone services.

Delay can be broken down into the following categories:

- Call setup delay (also called post-dialing delay);
- Propagation delay;
- Speech processing (compression/decompression) delay;
- Switching and queuing delay (packet and circuit-switched);
- Delay-induced impairments (echo).

Each is reviewed in the following paragraphs.

Call setup delay is the time between when the dialing party completes entering the number of the called party and the called telephone instrument rings and is answered (assuming no additional delay for a slow response at the distant end). Since public telephony, whether using satellite or terrestrial access, requires that a connection be established before user information can be transferred, this is extremely important to the acceptability of the service. Figure 4.8 illustrates the steps involved and the following equation suggests how these contributors can be summed [7]:

$$T_{tot} = 3T_s + T_i + 3E(T_p) + E(W)$$

Where T_s is a constant (deterministic) delay assumed as for the time before a dial tone is delivered to the caller, the time for a ringing message to originate from the called end, and the time for the answer indication to be given to the caller, T_i is a constant time for a connection request to traverse the network to the called end, $E(T_p)$ is the expected value (statistical average) of the switch processing time at nodes A and B, and $E(W)$ is the expected value of the waiting time for the call request message to reach switch node B.

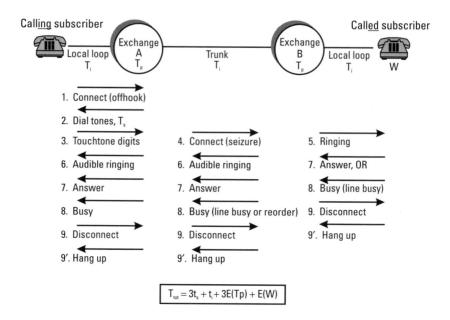

Figure 4.8 Telephone call signaling sequence, indicating transmission nodes, time duration, and delays.

This equation and Figure 4.8 are only meant to illustrate the components and how they combine to impact the call setup time. In a real case, the telephone engineer must evaluate in detail the delays at points of processing, switching, and transmission that affect call setup. This may entail actual measurements or, in the absence of a real system, computer simulation. Terrestrial fiber-optic networks that employ SS-7 produce low setup delay (less than 1 second), while the corresponding delay in cellular networks can be quite substantial (5 to 10 sec). Users who have experienced both have come to accept the difference in call setup delay performance.

Propagation delay is the time taken to travel over the cable or radio path. Presented in Chapter 3 and Figure 3.1, propagation delays for the LEO, MEO, and GEO altitudes are of the order of 7, 75, and 260 ms, respectively. LEO systems have the benefit of very brief single hop delay, which is comparable to delays experienced within telephone exchanges and multiplexing nodes. Multiple hops for LEO constellations would be quite acceptable, including intersatellite links. However, processing and switching onboard a satellite will add substantially to this delay, as discussed below. The propagation delay in MEO links is significantly greater, but is substantially below that of GEO satellites.

Propagation delay is only one component of the total budget, but for GEO systems it is usually dominant. Other contributors include compression and decompression processing and packet switching (if used). Experimental evaluation of MOS under various conditions of delay indicated that there is a cliff in the MOS rating curve at around 500 ms, a point where a significant fraction of telephone users find the circuit to be below 3.0. These scores are heavily influenced by the specific conditions of the call. For example, the best scores are always obtained for listening-only with steady-state signal quality. Two-way interactive conversation immediately brings out the delay characteristics and possibilities for speech clipping. However, subscribers involved in a mobile call with fading will tend to rate the same delay and quality with a higher MOS score (as much as half or a full point) because of a lower expectation.

Speech processing (compression/decompression) is a powerful technique to reduce the data rate necessary to transfer a voice channel. For heavy compression that reduces the bit rate to less than about 6 Kbps, the processing function will introduce a delay of its own. The amount of this delay, which is essentially constant and fixed for all telephone calls, results from the number of speech samples that are required and the actual processing time to perform compression and decompression.

Switching and queuing delay (packet and circuit-switched) is the consequence of internal node transfer and buffering of traffic within a network (satellite or terrestrial). While this is more of an issue for data communication in packet switched networks, it is still a natural consequence of having digital information transferred through time division switches. Also, as ATM is adopted throughout the global network infrastructure, processing and queuing delay will become a larger and larger factor in network design. An example of this type of consideration was already presented for call setup delay, but in this case, the delay affects all information flow after the connection is established.

Delay-induced impairment (echo) was a serious detriment for commercial satellite communications prior to the wide scale adoption of digital communications for international services. The problem was largely caused by a switching device called an echo suppressor, which was used in all long-distance circuits to block the reflection of echo. The half-second round-trip delay of GEO satellite links combined with poor echo suppression to produce service quality that was often unacceptable.

The solution to this problem was a digital device called the echo cancellor, which as the name suggests, actually removes the echo by canceling it out at the source. The hybrid at the distant end still returns the echo, but the echo cancellor introduces a negative-going replica of the echo to substantially remove it by numerical subtraction. When the circuit is first established, the echo cancellor goes through a training period to model the reflection path. The resulting attenuation of 40 dB is sufficient to make the circuit sound natural. In addition, hard echo suppression, which is digitally introduced, disables the path when only one person is actually speaking. The 40 dB of cancellation is really only required when both parties are speaking at the same time (called double-talking).

Echo cancellors are now standard in all long-distance circuits, whether on satellite or terrestrial links. This is important because all digital services involve the kinds of time delays previously discussed. For example, digital compression in terrestrial cellular networks introduces enough delay to demand that echo cancellation be provided within the digital cellular network (but not within the cellphone, since this is a four-wire telephone device). Likewise, the variable delay of the Internet requirement places an echo cancellation on VoIP services.

4.1.3 Traffic Engineering and Capacity

The design of traditional telephone networks was, for decades, viewed as a straightforward procedure using the analytical techniques derived by Agner Krarup Erlang (1878–1929), the first person to study this problem [8]. A Danish mathematician who worked for the Copenhagen Telephone Company, Erlang proved that Poisson's law of distribution governed the theoretical performance of circuit-switched networks. It is based on the principle that subscribers make calls independently of one another and these calls randomly arrive at the telephone exchange at an average arrival rate of λ calls per second. Mathematically, the probability of k arrivals in an interval of time T is given by:

$$p(k) = \frac{(\lambda T)^k e^{-\lambda T}}{k!}$$

This surprisingly simple formula tells us that the traffic demand can be specified exclusively by λ, the average arrival rate. What remains (to be discussed below) is how the switch and network respond to this demand. The individual demand per user is relatively small, perhaps only a few minutes during the busy hour. The traffic offered, A, to the exchange that routes calls over the channels is measured in Erlang (E), and defined as:

$$A = \text{call rate} \bullet \text{call duration}$$

For example, if the offered traffic is 0.5 calls per second and the average call duration is 30 seconds, then A = 15E.

The statistical property of telephone calls produces fluctuation in the loading of a particular trunk. If all N channels are in use, then the next call to arrive (the N + 1 call) will be blocked (e.g., the caller hears a "fast busy" tone). The telephone network could either put such calls in a queue awaiting available trunk capacity, or immediately reject the call (calls dropped). Since calls arrive randomly, N channels will carry fewer than N Erlangs of traffic, unless the network operator allows a very high frequency of blocked calls. Traffic loading depends on time of day, so we use the concept of the busy hour to form the basis of telephone network design.

A question might arise about the potential load offered by a single subscriber. Suppose that the typical subscriber makes one call during the busy hour and that this is two minutes in duration. Then the load for this subscriber is:

$$A = [1/3600] \bullet 120 = 0.0333 \text{ E or } 33.3 \text{ mE}$$

Thirty such users would produce the traffic load equivalent to one continuously occupied trunk.

A. K. Erlang derived the following mathematical relationship (the Erlang B equation for blocked calls dropped) to describe how this large community of potential callers will impact service on the trunks during the busy hour:

$$P_B = \frac{\dfrac{A^N}{N!}}{\displaystyle\sum_{n=0}^{N} \dfrac{A^n}{n!}}$$

where P_B is the probability that a call will be blocked because all trunks are busy.

For our simple example, 30 subscribers "offer" a traffic load of 1 E. If N = 3 channels in the trunk, this formula tells us that the probability is 6.25% (1 out of 16) that all channels are already occupied so that the next call is blocked, i.e., receives an "all trunks busy" signal from the switch. Normal telephone engineering practice would require P_B less than 1% (1 out of

100). A compliant design would provide five channels, for which $P_B = 0.3\%$. (It is impossible to have the precise solution at 1% because fractional channels are not physically realizable.)

The practical design approach in telephone network engineering is to predefine the value of P_B based on the quality of service that the network must provide to subscribers. In the PSTN, the accepted value is 1%, or one call blocked out of every 100 attempts. A graphical representation for a range of values of N and E is presented in Figure 4.9. We have chosen the Erlang B equation, above, which applies in the arbitrary case that all blocked calls are dropped, i.e., that the caller gives up if he or she cannot obtain the trunk for the call in question. The Erlang C equation is a refinement that allows for blocked calls to be held, that is, placed in a queue awaiting the next available trunk [9].

$$P_B = \frac{\dfrac{A^N N}{N!(N-1)}}{\displaystyle\sum_{n=0}^{N} \dfrac{A^n}{n!} + \dfrac{A^N N}{N!(N-1)}}$$

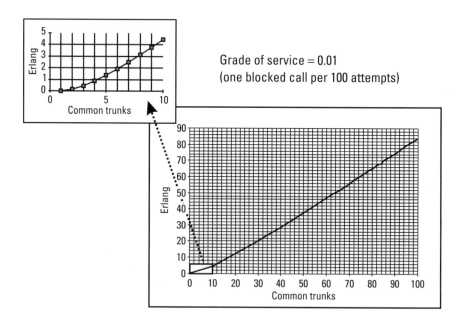

Grade of service = 0.01
(one blocked call per 100 attempts)

Figure 4.9 Traffic engineering design curves based on Erlang B equation (dropped calls cleared from network).

Whether blocked calls are cleared or held is largely a moot point, since neither of these elegant formulas is applied in practice. Rather, they provide a means to understand the statistical nature of telephone networks.

Estimating telephone traffic in a new network that has not been constructed, or worse yet for a service that does not yet even exist, is difficult, to say the least. The procedure seems to be to use either of A. K. Erlang's formulas as a rough starting point, and then to proceed with some kind of test case, possibly involving a pilot service on a testbed with live subjects. These data can then be used as an input to a computer simulation of part of the network under investigation. In this way, many scenarios can be tried and measured under a variety of conditions. The actual design to be installed should probably have excess trunk and switching capacity to alleviate blockages that can occur, and to reroute traffic as necessary during peak loading situations. It is hard to predict how new subscribers would react to a high level of blockage. For example, people in countries with low telephone density contend with such inadequate trunk facilities that one would speak of a call-completion rate as opposed to a call-blocking rate. More recently, first-generation cellular networks experienced overloading as their operators struggled to reach profitability during the early lean years. Subscribers on current fixed networks have become accustomed to having much better service levels available and so expectations are rising.

4.1.4 Service Management

Satellite telephone networks that deliver services to subscribers must be managed like a business, the most basic function being service management. One important aspect of service management is that the basic information about subscribers, their level of services, their usage records, and service bundles, must be generated within the network and made available to the operator of the network. The "hooks" for these functions must be introduced early in the development of ground segment components. Also, there could be requirements to pass such information back and forth across the interface to other service providers.

The basic arrangement of a service management system is illustrated in Figure 4.10 [10]. The following is a typical list of functions from the standpoint of telephone services that could be provided by a satellite communication ground segment:

- *User attributes and service bundle.* This identifies what services a particular user has requested and is entitled to employ. Along with this

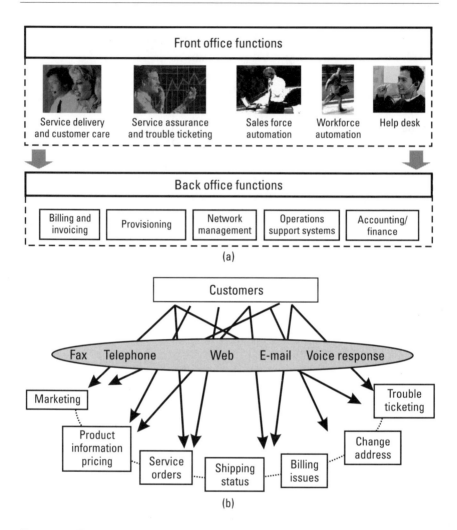

Figure 4.10 Telecommunications-based service management systems: (a) information architecture for front-office and back-office management of the service business (illustration courtesy of Clarify, Inc.); (b) customer interaction in the New Economy.

goes a pricing arrangement (referred to as rating), including the basic rate for service and usage-based pricing (e.g., price-per-minute during day and evening).

- *Accounting.* Accounting management deals with the generation and processing of information to measure usage of network resources (ground and possible space as well) and creates the call detail information fundamental to creating call records [11]. This is the basis of billing, discussed below.

- *Billing.* This uses information gathered as the user places calls, taking account of user attributes and the associated service bundle. Call records are gathered from the network and stored for monthly resolution. The bills can be printed and mailed, although there is greater interest in making billing data available online over the same network or the Internet. Payments must be collected and billing resolved so that the proper steps are taken to reduce delinquency and fraud. Some of the steps that can be used to improve payment performance are to employ prepaid calling cards or smart cards or to have all bills cleared with credit cards.

- *Mobility management.* Satellite-based networks typically allow users to move anywhere within the coverage area. Therefore, there should be a scheme for authenticating a user who activates his or her terminal from an arbitrary location and attempts to place calls. The classical scheme for doing this is the visitor location register (VLR) common in the GSM digital cellular standard [12].

- *Service provisioning and activation.* The service provider or reseller who acquires subscribers and puts them into the network in the first place needs a user-friendly system to activate the service on an individual basis. For a fixed-line solution, this also involves sending an installer out to the customer's location. Satellite services can permit customers to provision their own access by purchasing the necessary terminal from a distributor and subsequently doing a self-install at the business premise or home. When activated, the network should become more or less transparent to the subscriber, with he or she only having to make infrequent service inquiries regarding problems or new features. This function interfaces with the activation function described previously.

- *Fraud management.* Wireless networks have become the targets of fraud, the unauthorized use of the network without paying. The two paths where fraud is encountered are at the user terminal level, e.g., using a legitimate device available through the network to place illegitimate calls, or accessing the network over the radio link with a specially created or modified user device that intentionally bypasses the call accounting and billing systems. The former occurs when a user simply stops paying the bills when they arrive; this can be resolved by following up quickly on non-payers. The second abuse is usually more difficult to reduce because the perpetrators probably have inside information on the weaknesses of the system. Reducing this problem means either locating these criminals or blocking their access by closing the loopholes. Fraud management is fundamentally a game of cat and mouse, in which the network operator (the cat) must be extremely vigilant and aggressive to uncover the abuses and cut them off; the wrongdoer (the mouse) continues to find new and more intriguing ways around the system. Other security concerns include integrity of the data discussed in this section and limitation of damage due to violations of security processes.

The previous discussion was only meant as an introduction to the general topic of service management and should not be viewed as a comprehensive list of requirements for a given telephone ground segment. The requirements can only be determined once the specific nature of the services and network design has been determined.

The various computers, software systems, information networks, and supporting organizations that accomplish the above objectives represent a substantial investment and require considerable ongoing maintenance. In recent years, these functions have migrated to specialized software suites developed under the broad panoply of *customer relationship management* (CRM). In the telecommunications industry, the leader in providing comprehensive and well-integrated CRM solutions is Clarify, now part of Nortel Networks. One discovers quickly, however, that while CRM is very attractive for the operator-customer interface (including the provisioning aspects), it does not as yet address recording of calls and billing. These functions are complex in their own right and have, in the past, been addressed by switch manufacturers like Alcatel, Lucent, and Nortel. However, specialized billing software, with special features for flexible rating (e.g., complex billing algorithms to lure customers), have appeared on the market. One of the companies that provides billing solutions is Geneva, based in the United Kingdom. In the future, it is possible that, with the complexity of these software systems, operators will choose to outsource CRM and billing to major computer service firms like EDS and IBM.

4.1.5 Synchronization with Terrestrial Networks

We have already reviewed many aspects of the service and interface requirements of telephony, whether via an earth station or terrestrial PSTN facility.

Beyond these basic requirements, there are a host of other issues that can only be addressed once the local situation is understood and the appropriate scenarios described in detail. This requires a lot of research on the ground, including a thorough investigation of the existing PSTN standards and facilities. The fact that a given country or region has adopted an ITU, ETSI, or Bell System standard does not guarantee that new earth stations or switching equipment will perform properly.

One of the questions that often gets overlooked is that of synchronization of timing and clocks. Two fundamental approaches are applied.

- In plesiochronous mode, timing is inherently flexible to allow for a wide disparity in local timing. This is made possible by including dummy bits that can be removed or inserted at each node.

- Synchronous mode relies on close alignment of timing throughout the network. This demands high accuracy of individual clocks, of the order of 10^{-11}, in order to achieve maximum throughput without dummy bits. The benefit is provision of end-to-end clear channels for precise data and applications that demand this type of performance [13].

The goal of synchronous mode is to provide a stable reference that is traceable to a highly accurate primary reference source (PRS), such as a rubidium atomic clock, the U.S. National Institute of Standards and Technology (NIST) WWV radio broadcast, GPS, or the like. This timing can be further distributed over the satellite to all locations, a technique that can be successful as long as the local environment at the earth station is compatible. For example, the satellite network may be properly tied to the PRS, but the local switching office (which is independent of the satellite network) may not be so exact.

The slippage of clock timing can cause unexpected difficulties in the operation of certain types of equipment. Voice connections may not even notice a clock slip, which might cause nothing more than a click or dropout of a few milliseconds. Connections involving voice-band modems will not fare as well, possibly causing a session to end abruptly. A slip in a fax transmission could produce a vertical gap of a few millimeters. Encrypted services such as voice or data are very susceptible to slips, as these interrupt synchronization, a problem that impacts digitally compressed video as well. The solution may require forcing the newly introduced satellite network to follow the local clock, or vice versa.

4.1.6   Signaling in Telephone Networks

A second important consideration in the telephone interface has to do with the signaling system used for call setup, processing, and billing [14]. Telephone signaling has developed slowly over the decades because of the enormous established base of telephone exchanges in each country. The two basic types of signaling are supervisory and interregister (i.e., numerical). Supervisory signaling directs the process of initiation and teardown of the call. On the originating end, a signal is transmitted to "seize" the line, that is, to inform the other end that there is a demand for service on a particular line. The other end acknowledges the seize and holds the line open for the subsequent interregister signaling, which conveys the city or area code and local number. Table 4.5 provides a summary of techniques for each aspect of signaling on subscriber loops and trunks [15].

On most telephone networks, supervisory signals are energized through a separate pair of wires on transmission equipment, called the "ear" (E) and "mouth" (M) leads. The E lead accepts signaling input from the originating switch or transmission line, and the M lead transmits signaling to the next link in the chain or the terminating switch. With conventional transmission equipment, supervisory signaling is carried at a frequency of 3,825 Hz (3,700 Hz in North America), which is outside of the voice frequency band ("out of band") yet follows the same path as the associated voice channel. Subscribers cannot hear the tone because bandwidth filtering for

Table 4.5
Comparison of Loop and Trunk Signaling in Telephone Networks

Type of Signaling	Technique	Standard
Subscriber loop		
On-hook/off-hook	DC current	—
Dialing	Multifrequency code	—
Caller ID	PSK ~1,500 bps during ringing	—
Trunk, analog		
4-wire, supervisory	In-band	R1
6-wire, supervisory	Out-of-band, E&M	R2
Interregister	Multifrequency code	R1, R2
Common-channel signaling	Out-of-band, packet switched network, 64 Kbps	SS-7

300–3,400 Hz in the transmission system blocks it. Some older North American transmission systems use "in-band" supervisory signaling, also called single frequency (SF) signaling, with an audible tone frequency of 2.6 kHz. In this case, a narrow "notch" filter is used to remove the tone from the voice frequency bandwidth. AT&T discovered that telephone pirates, called "phone freaks," made free calls by using "blue boxes" that generate SF signals and interregister signals (discussed below) to fool long-distance switches into allowing nonpaying calls out of band through the network. Phone freaks are largely out of business by virtue of common channel signaling (SS-6 and SS-7), which do not recognize the tones coming from blue boxes.

During the 1960s and 1970s, the most common form of interregister signaling used the voice channel itself to pass the dialing digits, also called address information. In the in-band approach, each digit (0 through 12 or 0 through 16, depending on the system) is coded into a pair of audible tones, there being seven or eight different tones to be employed. You can listen to the combinations when you touch the keypad of a telephone. After the digits are received and interpreted by the switch at the other end, the call is completed to the distant party. When the party answers, the distant switch returns an appropriate supervisory signal to confirm the connection. Now the switching equipment can track the duration of the call for billing purposes. The final step occurs when one party hangs up, causing a "clear" supervisory signal to be transmitted. The switches take down the circuit connection so that another call can be connected through. Billing information is also recorded.

The predominant in-band signaling systems in North America and Europe are called R1 and R2, respectively, and were adopted as international standards by the ITU-T. The R2 system is also used in many countries of Asia and Africa. Variants of these systems have been developed to make them compatible with satellite links to bypass the "compelled" mode, which holds the tone on until a response has been received from the distant register. If a satellite link were introduced, this handshaking would add unacceptable time delays to set up a call. Semicompelled signaling was devised wherein tone pulses of fixed duration are used. The delay for call setup is reduced because the sending equipment waits only once per digit for acknowledgment before sending the next tone pulse.

International trunks over transoceanic cables impose additional complexity to the problem of interconnecting telephone exchanges. A form of semicompelled tone signaling, called ITU-T System Number 5 (SS-5), resulted from a joint development effort of AT&T Bell Laboratories and the

British Post Office. This system operates effectively over transoceanic cables that are equipped with time-assignment speech interpolation (TASI) equipment, and over satellite circuits as well. TASI is a circuit-multiplying technique that detects an active speaker at the input to the cable and switches the speech into an available channel. In digital telephone trunking, TASI is replaced by digital circuit multiplication equipment (DCME), which combines an advanced form of speech interpolation with voice compression through adaptive differential pulse code modulation (ADPCM). This can yield an effective gain in voice channels of from four to eight, depending on the number of voice channels available.

AT&T first introduced common channel interexchange signaling (CCIS) to combine the supervisory and interregister functions on a dedicated out-of-band data network backbone. Later, CCIS was adopted by the ITU-T as Signaling System Number 6 (SS-6), the precursor to the more advanced version called SS-7. Under this concept, the supervisory and interregister signaling information is stripped from the telephone circuit at the originating exchange and transmitted as a message over an entirely separate data communication link that runs in parallel with the voice network. The message format takes advantage of the principles of packet switching and processing, which can be made extremely reliable. Error-detection techniques such as cyclic redundancy checks and forward error correction allow the data communication network to yield messaging reliabilities of greater than 99.9999999%. The developers of SS-6 felt that this type of reliability was necessary because the distant exchange would not be in direct contact with the originating exchange when completing the call. Also, SS-6 does not interfere with the operation of DCME and is not seriously affected by satellite propagation delay.

The common channel signaling approach is integral to SS-7, whose datagram message structure is based on the Open Systems Interconnection model and is also the signaling system for ISDN. The physical channels transmit synchronous data at a rate of 64 kbps, raising the number of telephone channels within a given group by several multiples. At the link level, SS-7 employs high-level datalink control (HDLC), a robust protocol employing the "look back N" technique. The higher level protocols of SS-7 achieve reliable end-to-end message routing and facilitate the specific content structure of call processing messages. SS-7 includes ample capacity for new features for use by subscribers and by the telephone operators themselves, but this capacity must be activated by all players for the potential to be realized.

4.2   VSAT Data Networks

Very small aperture terminal (VSAT) satellite communication networks were introduced during the early 1980s in response to pent-up demand for a low-cost and reliable alternative to leased analog or digital telephone lines for business data communications. Consisting mostly of user-premises, small earth stations and a centralized hub main earth station, the VSAT network gained popularity by appealing to a segment of U.S. corporations, offering more effective service than that provided by analog telephone lines employing the 9,600 bps modems of the day. The first users in the United States included Wal-Mart, then an up-and-coming retailer that built its stores near small towns in rural America; Chevron, the first oil company to place VSATs at each of its company-owned filling stations; and Toyota of America, which wanted to enhance the quality of service provided to the first buyers of their new Lexus automobiles.

In reality, VSATs can be configured for essentially any kind of telecommunications service (not just data, but also telephone and video as well) and are suitable for installation on customer premises. The most visible part of the VSAT is its antenna, which typically is less than 3m in diameter. There is a trend to push size down to approximately 1m in an effort to reduce the cost and difficulty of installation on typical buildings. Ultimately, we wish to be able to locate VSATs at homes to encourage personal use and small office/home office (SOHO) applications. At the time of this writing, this type of equipment was not on the market, with the possible exception of receive-only data communication terminals like those provided for DirecPC and Cyberstar. The direction that we are going in is to allow two-way communication over the satellite, eliminating any terrestrial connection (unless the user wishes it for backup communication and to provide another type of application such as video on demand).

In the following sections, we provide a brief review of the basic arrangements that VSATs offer, as well as the types of applications that are currently supported. This should be viewed as the starting point for developing a specific ground segment and for defining requirements for the earth stations themselves.

4.2.1   VSAT Network Topologies

VSAT networks fall into two broad categories: the hub-based "star" VSAT network, which provides a star type of topology, and the "mesh" network, which allows connections between any pair of VSATs. Acting as the common relay point, the space segment should be fairly transparent to the operation of the network, unless we are speaking of a processor-based repeater design. The latter may offer hub characteristics without the overhead and added delay of a ground-based hub.

As with any satellite network, VSATs must employ a multiple access (MA) system that is both reliable and efficient. These methods include TDMA, which is the most common for VSAT networks, CDMA, which is gaining popularity in mobile networks, and FDMA, the original MA technique and still quite useful for mesh topologies. Hybrid networks that adopt two of these techniques have also appeared as a way to address some of the dynamic aspects of VSAT networks. For the mesh, all earth stations must transmit with sufficient power to reach any other earth station in the network (to permit "any to any" types of connections), while in the star network there can be an imbalance of power favoring the VSATs. This recognizes that very large networks of small earth stations (with antennas of around 1m and SSPA output power of less than 5 watts) benefit from using a common hub with a large antenna (6m to 10m) to be capable of receiving the low-power VSAT transmissions. This assumes the common analog bent-pipe type of repeater with linear (or near linear) gain characteristics. Later, we provide link budget examples to emphasize this particular point.

In the mesh network, we must still provide some form of centralized network control and management, since much of the communication will benefit from a demand assignment (DA) scheme of some type. This central control could use one of the mesh network earth stations to access the space segment. The purpose of allowing this is to permit any station to request a connection to any other and to obtain the necessary space segment bandwidth and power for the duration of this connection. The central control function is vital to the proper operation of the ground segment, which cannot function without the ability to assign and remove the necessary links and manage the network in a service sense. Channel capacity to be assigned can be in terms of therefore bandwidth within the transponder or baseband bandwidth in packet form.

The demand for VSAT applications has pushed the operating frequency higher into the Ku and Ka bands, where spectrum is more readily available than it is at C, S, or L bands. This means that VSAT links must accommodate significant rain attenuation, particularly in temperate and tropical regions of the planet. As discussed in Chapter 2, it is possible to obtain good link availability (i.e., at the 99.9% level) provided that there is adequate link margin. Another aspect of using Ku and Ka bands is that terrestrial interference from microwave stations will not be present in general and so VSATs can be operated from almost anywhere as long as the satellite (or satellites) can be seen without ground obstruction.

4.2.2 Computer Network Requirements

The first step in understanding where VSAT mesh and hub networks fit in is to examine how companies apply data communications to solve business problems. We include under "business problems" the kinds of applications used by nonbusinesses, including governments, educational institutions, and nonprofit organizations. All of us, then, can acquire the computer hardware and software resources to automate routine functions, access and process data, publish and distribute information, and more recently, engage in electronic commerce at the retail and interorganizational level. Some general examples defining the uses of VSATS are:

- As a replacement for an existing star or mesh network that serves an organization's data communication needs (this existing network may also provide other forms of communications, notably voice and video);

- To activate a new strategic information technology (IT) application to achieve a business goal of becoming more competitive in a particular market;

- As an integrator of wideband services, where the organization had been using a number of different networks and services, each addressing a different requirement;

- For rapid installation or expansion of an existing network, taking advantage of the unique features of satellite communication;

- As a platform for a corporate Intranet that addresses a wide variety of locations within a country or region, or even on a global basis. An Intranet is a closed computer network that uses the open architecture of the Internet, the Transmission Control Protocol/Internet Protocol (TCP/IP), but limits access and applications to those that support the internal needs of the organization.

Computer networks have long been viewed in terms of a layered architecture, typically following the seven layers of the Open Systems Interconnection (OSI) model [16]. Modern data communication theory and practice is literally built upon the concept of protocol layering, where the most basic transmission requirement is at the bottom, and more complex and sophisticated features are added one on top of another. As we move up the "stack," each layer obtains a service form from the layer immediately below. In this way, the details within the layer can be optimized for performance and isolated from the other layers. At the very top of the structure is the actual information processing application that is required in order for the network to carry out its function. These principles are summarized below with a definition of the specific layers (which are normally arranged with Layer 1 on the bottom, but listed here in ascending order, for clarity).

- *Layer 1—Physical.* Provides the mechanism for transmitting raw bits over the communication medium (cable, wireless, or satellite). It specifies the functional, electrical, and procedural characteristics such as signal timing, voltage levels, connector type, and use of pins. The familiar RS-232 connector definition is a good example of the physical layer. On satellite links, the physical layer is built into the modem and MA elements and must be developed to maintain a reliable point-to-point or point-to-multipoint transfer of data in the presence of link fades and noise-induced errors and interference.

- *Layer 2—Data Link.* Provides for the transfer of data between adjacent nodes or connection points, either by a dedicated point-to-point line (e.g., a T1 private line or a satellite duplex link) or a medium capable of shared bandwidth (e.g., an Ethernet cable or satellite TDMA channel). VSAT terminals and hubs must incorporate a basic protocol for maintaining reliable data transfer, often including a scheme for automatic retransmission of lost packets, done in a manner that is invisible to the higher-layer protocols by the computer application.

- *Layer 3—Network.* Responsible for routing information from end to end within the network, allowing for multiple data link paths. This may involve decisions as to the most effective route through the point-to-point links that comprise the network. A VSAT network may serve as one of these links and hence would have to interface properly with the network layer. Popular examples of the network layer are the X.25 protocol used internationally and the Internet Protocol (IP) that is employed in the majority of router-based private networks, as well as in the Internet itself.

- *Layer 4—Transport.* Provides another level of assurance that the information will properly traverse the network, from end user to end user. Two transport layer services are commonly available: *connectionless*, which transfers packets of data, one at a time; and *connection-oriented*, where a virtual circuit is first established before sending multiple packets that make up the entire conversation. The familiar TCP layer of TCP/IP provides a connection-oriented service to computer applications. Generally, the satellite system tries to use the terrestrial transport layer, although some form of spoofing (discussed later) may still be needed to improve throughput and reduce the average delay from end to end.

- *Layer 5—Session.* Somewhat more complicated than Layers 3 and 4, but provided to instill yet greater degrees of reliability and convenience of interface to applications. It manages the data exchange between computer systems in an orderly fashion to provide full duplex or half-duplex conversations. One important service is that of reestablishing the connection in the event that the transport layer is interrupted for some reason.

- *Layer 6—Presentation.* Provides syntactic and semantic services to the application layer above. What this means is that the presentation layer is inserted to resolve the complexities between transport/network layers and the more simplistic needs of the actual application that employs the network in the first place. Examples of syntax include encryption, compression, and numeric encoding (e.g., ASCII).

- *Layer 7—Application.* Includes the actual data communication applications that are common in open systems, such as file transfer, virtual terminal, electronic mail, and remote database access. We refer to these as applications because they include not only the protocol elements that support specific types of information but also features and facilities that ultimately interact with the end user. Most nonexpert users will not use the application layer directly, instead relying on specialized software within the computer to improve the interface and functionality. For example, most Internet surfers use the E-mail package supplied with the browser. This package, in turn, will engage Layer 7 electronic mail services (e.g., SMTP) to do the actual function of addressing, sending, and receiving message traffic.

The previous discussion was, of necessity, somewhat abstract as it relates to the logic of how networks are put together and interoperate.

Table 4.6 offers examples from terrestrial and satellite worlds to provide more of a practical feel for these aspects.

We see from Table 4.6 that satellite links differ significantly from terrestrial data communications at the lowest three layers: Physical, Link, and Network. This takes account of the substantially different technical environment of satellite links, particularly the delay, connectivity, and error-production mechanisms. Terrestrial links have their foibles regarding errors picked up along the route and delays encountered from links, nodes, and contention for these resources. The key point is that the widely used protocols, such as TCP/IP, X.25, and Frame Relay were designed for this environment. They can accommodate satellites as well, but usually in a nonoptimum way. What we find is that the protocol must be adapted to satellites or performance will probably suffer. An effort under the panoply of the Internet to create a facility called Multi-potocol Label Switching (MPLS) offers a long term solution for TCP/IP.

ATM is gaining in popularity for a wide range of applications and can support IP traffic as a backbone network. This allows integration of multiple information forms onto a single network architecture and achieves a higher degree of information transmission efficiency. An example of such an architecture, which had in years past been referred to as broadband ISDN (B-ISDN), is presented in Figure 4.11. In it, we see that a variety of user devices and access systems can connect to the ATM backbone using either a direct connection or a terminal adapter (TA) that converts the user protocol to ATM. Most popular at the time of this writing were TAs for LAN

Table 4.6
Protocols of the OSI Stack Layers, for Terrestrial and Satellite Implementations

Layer Definition	Terrestrial Examples	Satellite Examples
1. Physical layer	Fiber-optic cable, RS-232 connection, GSM cellular	QPSK modem-to-modem connection
2. Link layer	Ethernet, SONET, SDH	TDMA, ALOHA
3. Network layer	X.25, ATM, IP, Frame Relay	Proprietary protocols
4. Transport layer	TCP, SPX/IPX	Proprietary protocols
5. Session layer	Sockets, NetBIOS	Same as terrestrial
6. Presentation layer	ASN1	Same as terrestrial
7. Application layer	http, SMTP, X.400	Same as terrestrial

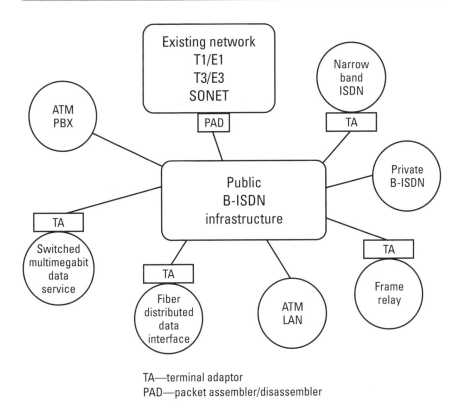

TA—terminal adaptor
PAD—packet assembler/disassembler

Figure 4.11 Application of existing and evolving networks to a public B-ISDN infrastructure.

protocols (e.g., Ethernet and Token Ring), IP (usually incorporated into the router), and Frame Relay. It should be kept in mind that whatever protocol is used on the input TA must also be used on the output TA (e.g., an Ethernet TA on one end must be matched by an Ethernet TA on the other).

The protocol layering of ATM is illustrated in Figure 4.12, indicating the ATM layer (with its 53-byte packet) at the middle. This amounts to a basic network layer that supports all types of services, such as constant bit rate (CBR) connections for voice and conventional video, and variable bit rate (VBR) for dynamic data communication involving IP and similar protocols on the end-user side. It is the function of the ATM adaptation layer (AAL), which resides just above the ATM layer, to perform the packet assembly and disassembly (PAD) functions that these user services require. At the physical layer we see a convergence sublayer to insert the 53-byte

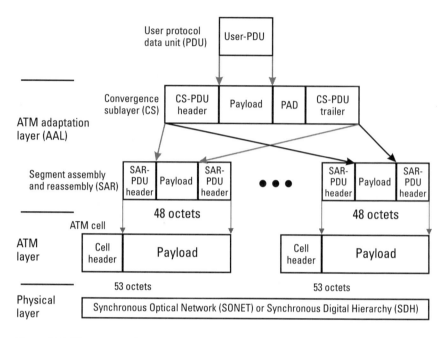

Figure 4.12 The protocol layers and cell structure of ATM.

asynchronous cells into the digital transmission scheme, which is to carry ATM-based data from point to point. Examples of convergence sublayer implementations include SONET/SDH, T1/E1, and Ethernet.

While ATM is growing rapidly at the time of this writing, its use over satellite links is limited—mostly to point-to-point connections between countries over INTELSAT, PanAmSat, or another international or regional satellite system. However, ATM plays a role in broadband satellite networks, particularly with digital processing repeaters. Fixed-length cell transmission will integrate the ground segment with telecommunication networks as public ATM gains acceptance. Readers should keep in mind that the 53-byte approach of ATM is rather simple; the complexity comes with the adaptation of ATM to the wide variety of higher-layer protocols and applications that exist and will appear in coming years. Considerable information on the evolution of ATM, including current and pending standards, can be obtained from the ATM Forum (www.atmforum.com), an industry-supported body that assists both suppliers and users of ATM technology.

Another commonly used protocol and network architecture is Frame Relay, which is based on the Link Access Protocol—Balanced (LAP-B) international standard. Frame Relay was introduced in the early 1990s to exploit the low-cost bandwidth made available on the newly developed fiber-optic networks in North America and Europe. Major IT vendors like IBM, AT&T, and Cisco Systems made the hardware and software available to facilitate introduction of Frame Relay systems and services on a national and international basis. In addition, the ATM Forum promoted implementation of Frame Relay over ATM to provide for the interoperability and longevity of both of these standards.

Figure 4.13 illustrates a typical Frame Relay network supporting a business with several major locations. The advantage of this approach is that the user pays for the access speed needed at each location (they need not be equal but rather are tailored to local requirements) and for the total aggregate data throughput of the network (termed the *committed information rate,* or CIR). This reduces costs substantially as compared with using dedicated leased lines between these locations. Additional information on Frame Relay can be obtained by consulting the Frame Relay Forum (www.frforum.com).

Throughout the discussion of efficient networks using ATM and Frame Relay, the basic premise remains that IP will dominate data applications. Much has been written about the design, features, and application of IP [17]. With its variable-length packets, ability to offer networking solutions for a wide range of protocols, and support through innovative

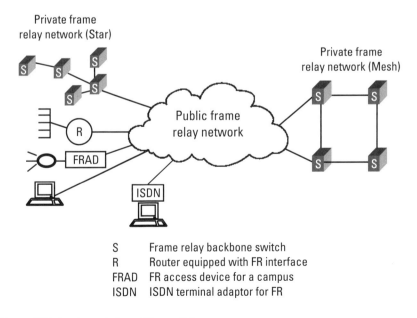

S	Frame relay backbone switch
R	Router equipped with FR interface
FRAD	FR access device for a campus
ISDN	ISDN terminal adaptor for FR

Figure 4.13 Implementation of Frame Relay.

companies like Cisco Systems, Sun Microsystems, Netscape (now part of AOL), and Microsoft, IP has surpassed the critical mass stage. There are some compelling reasons for this. First, IP is the foundation of the Internet and has contributed to its popularity. IP and the applications that employ it are extremely popular and pervasive, as seen in the success of PC-based networks in North America and western Europe and the spread of this technology to other areas of the world. Second, the cost of using IP has dropped to the point that it can compete with any proprietary or more sophisticated open protocol (witness the failure of the OSI suite of protocols, which offer much greater flexibility and security). Third, all computer hardware and software supports it, eliminating the compatibility problems that plagued the widespread use of computer networks in past decades. That IP by itself is not the final answer is affirmed by the long lifetime of ATM and Frame Relay. Other hybrid options include IP Switching using ATM as the underlying protocol, IP Tunneling to create a virtual private network (VPN) carrying a different protocol such as Novell Netware's IPX/SPX over the Internet, and VoIP, which permits conventional phone conversations to be made through the Internet.

This was a brief review of requirements related to data communication networks that may be served through the ground segment. The field is evolving daily, and we can only attempt to foresee basic trends and general approaches. Taking this approach, we now move into a discussion of the specific architectures that are in use at the time of this writing. We can anticipate that new arrangements, including the use of onboard processing repeaters, will add functionality to data communication ground segments, which should increase their penetration into global telecommunication markets.

4.2.3   Hub-Based VSAT Networks

The strength of the VSAT hub network, illustrated in Figure 4.14, lies in its ability to support low-cost VSATs with small antennas and low-power SSPA transmitters. The data communications services are centralized in topology to connect to a host computer located at a company or organization headquarters. VSATs are placed at remote locations that need data, voice, and video services from the hub; there is, by design, little or no communication between and among remotes. Rather, the vast majority of traffic is between VSAT and hub, reflecting that the central location has the data (or requires the data) that the remote computers provide (or display/store).

Interface requirements for VSATs are satisfied by the functionality of the indoor unit and internal port cards used to interface on the user side.

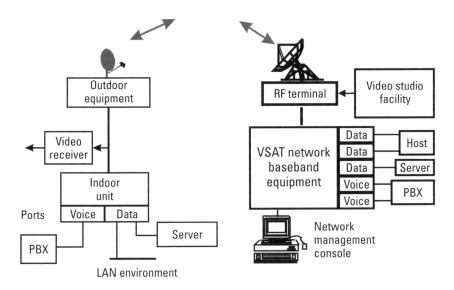

Figure 4.14 Data/voice/video network using a central hub earth station and remote VSAT terminal.

Customer equipment like PCs, a PBX, fax machine, or an IP router within a LAN or WAN environment are typical attachments. Consequently, the VSAT port is designed to accommodate a range of potential connections and to provide the needed spoofing of standard protocols such as TCP/IP and X.25. Within the indoor unit are the multiple access controller and custom modem for the space link. These components form the basic link and physical layers and are proprietary to the manufacturer of the system. Operators must fully understand what protocols need to be supported and that the VSAT network meets this need. Similarly, the hub must have appropriate protocol support for the host computer, which often is connected to it with a terrestrial data line of some distance. Operation of the end-to-end system is centralized to satisfy the demands of user applications and the layers of the protocol stack.

The use of the space segment at Ku band is indicated by the link budget for a typical outroute and inroute in Table 4.7. Hub characteristics include a 6-m transmit/receive parabolic antenna, providing an outroute EIRP of 68.9 dBW and a downlink G/T of 34.7 dB/K. The VSAT transmits its inroute with an EIRP of 48.3 dBW, using a 5-watt SSPA connected to a 1.2-m antenna (the actual power at the antenna transmit port is 3.5 watts, after waveguide and coupling losses). The receive G/T of the VSAT is

Table 4.7

Link Characteristics and Link Budget for a Typical Ku-band Hub-based VSAT Network

	Outroute	Inroute	Units
Uplink			
Uplink antenna	6.0	1.2	meters
Antenna gain	56.9	42.9	dBi
Transmit power	12.0	5.4	dBW
Uplink EIRP	68.9	48.3	dBW
Path losses	207.0	207.0	dB
Spacecraft G/T	2.0	2.0	dB/K
C/T_{up}	−136.1	−156.7	dBW/K
K	−228.6	−228.6	dB/W/Hz
Bandwidth	500.0	100.0	kHz
IndB	57.0	50.0	dB(Hz)
C/N_{up}	35.5	21.9	dB
Downlink			
Downlink EIRP	25.0	4.4	dBW
Path losses	205.7	205.7	dB
Downlink antenna	1.2	6.0	meters
Antenna gain	41.5	55.5	dBi
System noise temp	120.0	120.0	K
Earth station G/T	20.7	34.7	dB/K
C/T_{dn}	−160.0	−166.6	dBW/K
Bandwidth	57.0	50.0	dB(Hz)
C/N_{dn}	11.7	12.0	dB
Combined Link			
C/N_{th}	11.6	11.6	dB
C/IM	18.0	18.0	dB
C/I	15.0	15.0	dB
C/N_{tot}	9.3	9.3	dB
Threshold	7.0	7.0	dB
Margin	2.3	2.3	dB

20.7 dB/K, which is obtained from a low-cost 80K LNB. Losses and gains are calculated in accordance with the procedure discussed in Chapter 2. The link budget indicates that thermal noise, transponder intermodulation, and RF interference from adjacent satellites combine to produce an overall C/N of 9.3 dB in the outroute and inroute directions, which are both about 2 dB above threshold. This should provide adequate margin for rain fade in most temperate climates.

The outroute and inroute in this example of a typical hub VSAT network are unbalanced because their respective data rates are 512 kbps and 128 kbps, respectively. This means that there would have to be multiple inroutes to create a balance of information transfer between the hub and each VSAT. The exact quantity of inroutes will depend on the amount of inbound data throughput to the hub (which would be more or less than the outbound data, depending on the application). Techniques for evaluating and specifying these components can be found in [18]. The link budget does not include a reserve of hub power for uplink power control, which may be used to compensate for uplink fade (which affects reception at all of the VSATs). Satellite capacity for this network would be reserved in a transponder that contains multiple carriers, e.g., single channel per carrier (SCPC) or, alternatively, a partial transponder.

4.2.4 Mesh VSAT Networks

Mesh topology achieves direct point-to-point connection between pairs of locations with minimum propagation delay and satellite capacity utilization (in terms of the number of links if not power) with the mesh type of topology. For many ground segments, the mesh topology is also attractive because of its flexibility for connecting arbitrary pairs of earth stations. The same arrangement may allow any single earth station to broadcast data to all others. The mesh has been implemented in past years using any of the available MA techniques. As discussed in Chapter 1, mesh networks involving large earth stations have been in use since the 1970s for demand-assigned telephone services. Telephone service imposes severe requirements on the design of the network in terms of call setup times, blocking of calls, bit rate, and service management.

There are fundamentally three classes of connectivity requirements for mesh networks using VSATs. These are:

- *Permanently assigned* (dedicated bandwidth) links for CBR data transfer. This is analogous to a dedicated data line or digital private line using some multiple of 64 kbps, including T1 and E1 services. Higher data rates from DS3 and OC3 are feasible but would put the links out of the class of VSATs under discussion. An example of the latter type of application is to allow direct connection from a remote location to an Internet service provider (ISP), in situations where terrestrial lines of appropriate bandwidth are either not available or too expensive.

- *Demand-assigned* (circuit switched) links between locations for CBR data transfer. Again, the link must not require greater EIRP and G/T from the VSATs than is feasible. A telephony type of VSAT can also offer connection-oriented data services.

- *Dynamic allocation* of bandwidth (packet switched), allowing VBR services on a mesh network basis. This is the same type of facility provided by hub VSAT networks, with the exception that transmissions can be received by any earth station. With the introduction of Ka-band payloads and onboard processing, VSATs of this type will provide a wide range of services using a high-speed protocol such as ATM.

The attractiveness of VSAT mesh networks will be a factor in the growth of ground segments and satellite systems in coming years. At the time of this writing, non-GEO mobile satellite systems had just been introduced. These systems are based on extremely low-cost ultrasmall earth stations. While they provide many special benefits of true portability and operation literally anywhere, they are limited as to data transfer. This is where fixed VSATs hold the greatest promise for medium and high data rate services.

References

[1]   Freeman, Roger L., *Telecommunications Transmission Handbook*, 4th ed., New York: Wiley, 1998, p. 50.

[2]   The Communications Handbook, CRC Press, with Permission. 1996 (their Figure 27 [Fig.4.3]; their Figure 27.4 [Fig. 404]; their Figure 27.5 [Fig. 4.5]).

[3]   ADSL Forum, www.adsl.com/pressroom/dsl_flavors.html.

[4]   Rathgeb, Erwin P., "Integrated Services Digital Network (ISDN) and Broadband (B-ISDN)," *The Communications Handbook*, ed. Jerry D. Gibson, Boca Raton, FL: IEEE Press/CRC Press, 1997, p. 581.

[5] "Telephone Transmission Quality, Series P Recommendations," *CCITT,* Vol. 5, 9th Plenary Assembly, Melbourne (location of meeting), 14–25 November 1988; International Telecommunication Union, Geneva (home of ITU), 1989.

[6] Elbert, Bruce R., *The Satellite Communication Applications Handbook,* Norwood, MA: Artech House, 1997, p. 400.

[7] Schwartz, Mischa, *Telecommunications Networks: Protocols, Modeling and Analysis,* Reading, MA: Addison-Wesley, 1987.

[8] Erlang, A. K., "The Theory of Probabilities and Telephone Conversations," *Nyt Tidsskrift for Matematik B,* Vol. 20, 1909.

[9] Boucher, James R., *Voice Teletraffic Systems Engineering,* Dedham, MA: Artech House, 1988, p. 46.

[10] Bartholomew, Martin F., *Successful Business Strategies Using Telecommunications Services,* Norwood, MA: Artech House, 1997.

[11] Aidarous, Salah, and Thomas Plevyak, *Telecommunications Network Management: Technologies and Implementations,* Piscataway, NJ: IEEE Press, 1998, p. 165.

[12] Macario, R. C. V., *Modern Personal Radio Systems,* London: Institution of Electrical Engineers, 1996, p. 137.

[13] Freeman, *Telecommunications Transmission Handbook,* p. 233.

[14] Elbert, Bruce R., *International Telecommunication Management,* Norwood: Artech House, 1990.

[15] Noll, A. Michael, *Introduction to Telephones and Telephone Systems,* 3d ed., Norwood, MA: Artech House, 1999, p. 203.

[16] Elbert, Bruce R., and Bobby Martyna, *Client/Server Computing: Architecture, Applications and Distributed Systems Management,* Norwood, MA: Artech House, 1994.

[17] Wilder, Floyd, *A Guide to the TCP/IP Protocol Suite,* Norwood, MA: Artech House, 1998.

[18] Elbert, *The Satellite Communication Applications Handbook,* Norwood, MA: Artech House, 1997. p. 301.

5

One-Way (Broadcast) Service Requirements

The point-to-multipoint property of GEO satellites is vital to the entertainment industry and is gaining acceptance for a number of data distribution applications as well. By broadcasting the same signal across a wide area, a GEO satellite provides an excellent distribution point for an entire nation or region, and to cover the globe requires a total of only three satellites. This has made television services like CNN International, the BBC World Service, MTV, HBO, and CCTV into widely recognized brands for news and entertainment. Such a use is perhaps the most viable "killer app" for satellites and will likely remain prominent for decades to come.

From a ground segment standpoint, these broadcast-like applications use the point-to-multipoint capability of GEO satellites and thus rely on a centralized uplink facility to originate much of the programming. The typical broadcast center has the capability to both originate programming and relay material that is acquired over inbound terrestrial or satellite links. Since we rely heavily on digital compression and service multiplexing, the broadcast center is a facility that compresses, modulates, and uplinks multiple video, audio, and data channels for distribution over the coverage area. The user terminals are consumer items that can be purchased for a few hundred dollars (often given away to entice subscription) and can be operated via a convenient remote.

Other information forms have been distributed by satellite, and we anticipate that these will also rise in importance, although possibly not to the same business level as video. Included are data broadcasting, used to distribute character-based information, images, and other file types to vehicles and fixed locations, and audio broadcasting. For the mobile implementations of these services, it would not be necessary to employ GEO satellites, as some systems are expected to use a highly elliptical orbit (HEO) similar to that used within the Molnya system of Russia. Equipment for the reception of mobile broadcasts is expected to be included with normal car radios or attached like a portable CD player. The following sections review the requirements for these services as rendered by satellites.

5.1 Video Broadcasting

The ability of a satellite to serve a particular TV market is determined by the coverage area footprint. Thus, the Galaxy satellites are optimized to serve the United States, while Astra satellites cover Europe. The TV signals themselves must be consistent with the technical standards and content characteristics of the region served, aiming for one or more particular user segments. The important markets include network broadcasting to local over-the-air TV stations, cable TV (CATV) systems, and direct-to-home (DTH) subscribers who own their own dishes.

Satellite distribution of entertainment TV covers three application areas: local broadcasting (i.e., reaching viewers via VHF and UHF channels), CATV, and DTH digital broadcasting. In actuality, a given broadcast uplink services only one of these classes because of the targeted nature of the programming and the transmission format (analog or digital). Of fundamental importance are the standards used in the creation, organization, and distribution of the programming product. In the analog domain, the same format is used during each stage of preparation and delivery. This imposes tight specifications on the transmission performance of the channel, particularly the video signal-to-noise ratio (S/N) and impairments such as differential phase and differential gain that distort the picture. Digital formats like CCIR 601 are more tolerant of noise and distortion since they can be assessed in terms of the bit error rate alone. During the preparation of the product, the general view is that no impairment should be introduced and the highest data rate possible should be used. Distribution of the signal to the consumer occurs with the lowest data rate that is consistent with an adequate perceived quality using an MPEG standard (discussed next).

Network broadcast includes various forms of entertainment transmitted at VHF and UHF to home television receivers. Once carried out as a local service, network broadcasting is now a national or international medium. The uplinks to transmit programming may be right at the originating studio or could be connected to it through a long-haul fiber or satellite link (this is often required for multiregional broadcasts). Businesses also exploit the live-action nature of television through private broadcasting and two-way interactive video teleconferencing. In the case of the latter, the medium provides the means to engage in an enhanced form of interpersonal communication where the content is created in real time.

Cable and DTH TV are intended to be subscription services (meaning that the viewer pays), and use conditional access to perform functions such as:

- Allowing certain types of set-top boxes with appropriate smart card or other device technology to reduce the possibility of piracy;

- Providing program content with an effective system of scrambling to enable a single receiver to receive different services;

- Defining a standard interface specification for other devices such as digital receivers and PCs;

- Providing facilities for tracking down pirates;

- Arranging for licensing of the technology to manufacturers and providing an associated code of conduct for them to follow;

- Potentially allowing for the transfer of control at interfaces, such as between the satellite network and local cable TV system;

- Allowing for receiver mobility for when the owner wishes to relocate or install the receiver on or in a vehicle of some type.

Program distribution systems throughout the industry employ all or some of the architecture shown in Figure 5.1. There are many options as to how this architecture can be implemented, but the most economical approach is to combine these elements into a single broadcast center. This is unique to the satellite industry, where one organization can create, distribute, and sell a product from one location. For extremely large operations in DTH, for example, it could be prudent to achieve redundancy with an alternate broadcast site. Network broadcasters like CBS, Fox, and DIRECTV exploit the benefits of this arrangement. The functions of the broadcast

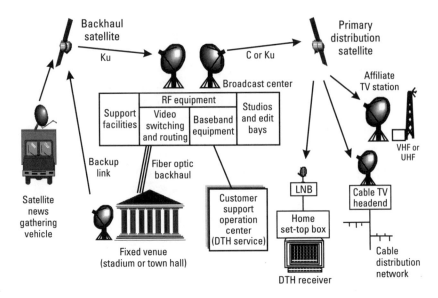

Figure 5.1 Combined satellite TV network, used to acquire real-time programming and distribute to TV stations, cable TV systems, and DTH receivers.

center depend on the specific application but include all or most of the following:

- Program origination in a studio;
- Display of prerecorded material, on film or tape;
- Program delivery by terrestrial links from remote studios and other venues such as a football stadium;
- Program delivery by satellite links from syndication sources and other video channels (e.g., cable TV networks);
- Reception of material using electronic news gathering (terrestrial microwave);
- Reception of material using Ku-band satellite news gathering (SNG);
- Editing of these various inputs into the actual program to be broadcast;
- Relay of programming captured from satellite links;
- Transmission from the studio to a satellite uplink (either C or Ku band);

- Transmission from the studio to a local broadcast transmitter site (for local broadcasting, if appropriate);
- Control and switching of input and output video and audio signals, to prepare program material for recording and distribution.

5.1.1 MPEG-2 Compression

Digital compression by way of the MPEG series of standards plays an important role in modern video and audio transmission. Its principal benefits are as follows:

- *Reduced transmission bandwidth,* saving space segment costs and reducing the amount of power needed to transmit an acceptable signal.
- *More channels available per satellite,* greatly increasing the variety of programming available at a given orbit position. This promotes new services like impulse PPV (Pay-per-View) and home education, and makes it feasible to expand a programming service through tailoring (e.g., packaging several different "feeds" of the same material with different advertising or localized content) and multiplexing (i.e., sending the same channel at several different times).
- *Common format,* used for satellite DTH, cable TV, and terrestrial broadcasting, based either on the U.S. DTV standard or the European DVB standard. This also provides a base for HDTV in the digital mode, because the number of bits per second of a compressed HDTV signal is less than what was previously required for a broadcast-quality conventional TV signal.

Table 5.1 gives an indication of the relationship between bit rate and application in commercial broadcasting. Perfect video reproduction of analog TV standards (NTSC, PAL, and SECAM) is achieved with rates of 90 Mbps or greater. Typical viewers cannot usually tell that anything is impaired when the signal is compressed to a rate of 20 Mbps. Below this value, it becomes subjective. For movies, a rate as low as 1.5 Mbps, the standard T1 in North America and Japan, is sufficient. However, for any live-action sports, at least 5 Mbps will be needed. In time, the performance of current compression systems and standards will allow improvements in motion performance so that 1.5 Mbps may meet the needs of all types of content. Use of multiple channels per carrier (MCPC) offers greater compression of these rates when combined with statistical multiplexing.

Table 5.1

Typical Data Rate Requirements for Production and Distribution of Network TV Signals

Purpose	Data Rate, Mbps
Program production	250 (nominal, using CCIR 601)
Backhaul transmission	34–45
MPEG-2 delivery	3–6

Compression systems that operate at 45 Mbps or greater are designed to transfer the signal without permanent reduction of resolution and motion quality. They are said to be "lossless" in that the output of the decoder is identical to the input to the encoder. In contrast, operation below about 20 Mbps is "lossy" in that it introduces a change in the video information that cannot be recovered at the receiving end. Lossy compression can produce a picture of excellent quality from the viewer's point of view at data rates above about 5 Mbps. There is an intermediate position called quasi-lossless, wherein a lossy compression service is augmented with the parallel transmission of an error signal that contains correction data to re-create a lossless image at a compatible receiver. Application of lossy transmission with reduced picture quality may be attractive, since it can reduce transmission costs (or allow the user to employ an existing communications system such as a VSAT network).

Transform coding is the most popular technique for reducing the number of bits required to represent a digital image. The basic idea is to replace the actual image data with a mathematically derived set of parameters that can uniquely specify the original information. The parameters, which are the coefficients of a mathematical transform of the data, require less transmitted bits to be stored or sent than the original image data itself, because they can be compressed further. Over the years, the discrete cosine transform (DCT) has proven to be the most popular mathematical procedure, and is now part of the JPEG and MPEG series of standards. The mathematical formulation of the DCT in both the forward and inverse directions is as follows:

For the forward transform used to encode the waveform:

$$F(u,v) = \frac{4c(u)c(v)}{N^2} \sum_{i=1}^{N-1}\sum_{j=0}^{N-1} f(i,j) \cos\frac{(2i+1)u\pi}{2N} \cos\frac{(2j+1)v\pi}{2N}$$

and for the inverse transform to decode the waveform:

$$f(i,j) = \sum_{i=1}^{N-1}\sum_{j=0}^{N-1} f(i,j)\cos\frac{(2i+1)u\pi}{2N}\cos\frac{(2j+1)v\pi}{2N}$$

where

$$c(w) = 1/52 \text{ for } w = 0,$$
$$1 \text{ for } w = 1, 2, \ldots, N-1$$

The Motion Picture Experts Group, also affiliated with ITU-T and ISO, has provided the MPEG series of standards for motion pictures and video, with the following desirable features:

- MPEG supports a wide variety of picture formats with a very flexible encoding and transmission structure.

- It allows the video service to use a range of data rates to handle multiple video channels on the same transmission stream and to allow this multiplexing to be adaptive to the source content (e.g., statistical multiplexing).

- Algorithms can be implemented in hardware to minimize coding and decoding delay (typically less than 150 ms at the time of this writing).

- Developers can include encryption and decryption to comply with content restrictions and the needs for business integrity.

- Provision can be made for an effective system of error protection to allow operation on a variety of transmission media such as satellite and local microwave links. (An excellent example is the DVB standard, discussed at the end of this chapter.)

- The compression is adaptable to various storage and transport methods, allowing content to be accessed by conventional TV receivers and home PCs.

- Transcoding will permit conversion between compression formats, since MPEG is widely popular.

- MPEG-processed videos can be edited by systems that support the standard.

- The standard will most probably have a long lifetime, since it can adapt to improvements in compression algorithms, VLSI technology, motion compensation, and the like.

An important contribution of MPEG-2 is its integrated transport mechanism for multiplexing video, audio, and other data through packet generation and time division multiplexing. It is an extended version (or superset) of MPEG-1, and is designed to be backward compatible with it. The definition of the bit structure, called the syntax, includes a bit stream, a set of coding algorithms, and a multiplexing format to combine video, audio, and data. New coding features were added to improve functionality and enhance quality, in the conventional video environment of interlaced scanning and constrained bandwidth. The system is scalable in that a variety of forms of lossless and lossy transmission are possible, along with the ability to support HDTV.

Because delivery systems and applications differ widely, MPEG-2 provides a variety of formats and services within the syntax and structure, called profiles and levels [1]. A profile is a subset of the MPEG bit stream that supports different classes of applications. A higher profile means more complexity in terms of the compression algorithm. A level defines a set of constraints on parameters in the bit stream profile. The higher the level, the greater the need for hardware/software processing speed. The profiles and levels are defined in Tables 5.2 and 5.3, respectively. Table 5.2 begins at the lowest profile, called SIMPLE, which corresponds to the minimum set of capabilities. Going down the table adds functionality at the expense of increasing complexity.

The tools and formats of MPEG-2 allow as many as 20 different combinations, which the standard calls convergence points. As with any such standard, not every combination is either useful or viable. At the time of this writing, the main profile and main level represent the convergence point of all practical implementations. This is the case in North America with the various systems already in use as well as with the European DVB standard, to be discussed.

The MAIN profile is the current baseline for MPEG-2 applications and has been implemented in a number of DTH systems. As suggested in Table 5.2, this profile does not include scalability tools and therefore is a point of downward compatibility from the higher levels that provide scalability.

The basic facilities of the lower layers of the MPEG protocol blend with the upper layers provided by DVB or a proprietary alternative such as that used by DIRECTV in the United States. The scheme for representing

Table 5.2
Profiles and the Associated Algorithms for the MPEG-2 Standard

Profile	Algorithms
SIMPLE	Provides the fewest tools but supports the 4:2:0 YUV representation of the video signal.
MAIN	Starts with SIMPLE and adds bidirectional prediction to give better quality for the same bit rate. It is backward compatible with SIMPLE as well.
SPATIAL Scalable	This and the profile that follows include tools to add signal quality enhancements. By "spatial" it is meant that the added signal complexity allows the receiver to improve resolution for the same bit rate. There would be an impact on the receiver in terms of complexity and hence cost. It can also be a means to add HDTV service on top of conventional resolution (i.e., only the appropriately designed receivers can interpret and display the added HDTV information).
SNR Scalable	The added signal information and receiver complexity improve viewable single-to-noise ratio (S/N). It provides graceful degradation of the video quality when the error rate increases.
HIGH	This includes the previous profiles plus the ability to code line-simultaneous color-difference signals. It is intended for applications where quality is of the utmost importance and where there is no constraint on bit rate (such as within a studio or over a dedicated fiber-optic link).

Table 5.3
Levels and Associated Parameters for the MPEG-2 Standard
(The bandwidth [BW] indicated in the last column is based on direct QPSK modulation
of the bit stream and does not include forward error correction coding)

Level	Samples/line	Lines/frame	Frames/s	Mbps	BW (MHz)
HIGH	1920	1152	60	80	40
HIGH 1440	1440	1152	60	60	30
MAIN	720	576	30	15	7.5
LOW	352	288	30	4	2.0

this is shown in Figure 5.2. Another dimension of MPEG-2 allows greater bit rate and facilities for a range of qualities from low definition to HDTV. There is an obvious impact on the bit rate and bandwidth. These features are reviewed in Table 5.3.

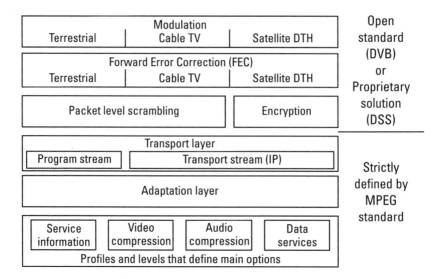

Figure 5.2 Structure of MPEG-2, indicating its protocol stack and support for satellite and terrestrial broadcasting networks (reference: http://www-it.et.tudelftnl).

These levels are associated with the format of the originating source video signal and provide a variety of potential qualities for the application. It ranges from limited definition and the associated low data rate all the way up to the full capability of HDTV. Another feature of the standard is that it permits the normal TV aspect ratio of 4:3 as well as the more appealing movie screen or HDTV aspect ratio of 16:9. This particular part of MPEG-2 covers the base input and does not consider the degree of compression afforded by the profiles covered in Table 5.2. The basis of each of the levels is as follows:

- Low level is an input format that is only one quarter of the picture defined in ITU-R Recommendation 601.
- Main level has the full 601 input frame format.
- High-1440 level is the HDTV format with 1,440 samples per line.
- High level is an even better HDTV format with 1,920 samples per line.

The other important element of MPEG-2 is the provision of stereo audio and data. The audio compression system employed in MPEG-2 is based on the European MUSICAM standard as modified by other algorithms. It is a lossy compression scheme that draws from techniques already within MPEG. Like ADPCM, it transmits only changes and throws away data that the human ear cannot hear. This information is processed and time division multiplexed with the encoded video to produce a combined bit stream that complies with the standard syntax. This is important because it allows receivers designed and made by different manufacturers to be able to properly segment the information. However, what the receiver actually does with the information depends on the features of the particular unit.

5.1.2 Digital Video Broadcasting (DVB) Standard

The DVB standard is an outgrowth of work in Europe of the European Broadcasting Union (EBU) and the European Telecommunications Standards Institute (ETSI) [2]. The overall philosophy behind DVB is to implement a general technical solution to the demands of applications like cable and DTH TV, including:

- Information containers to carry flexible combinations of MPEG-2 video, audio, and data;

- A multiplexing system to implement a common MPEG-2 transport stream (TS);

- A common service information (SI) system that provides data for the onscreen electronic program guide (EPG);

- A common first-level Reed-Solomon (RS) forward error correction system (this improves reception by providing a low error rate to the decoded data, even in the presence of partial link fades);

- Modulation and additional channel coding systems, as required, to meet the requirements of different transmission media (including FSS and BSS satellite delivery systems, terrestrial microwave distribution, conventional broadcasting, and cable TV);

- A common scrambling system;

- A common conditional access interface (to control the operation of the receiver and assure satisfactory operation of the delivery system as a business).

The DVB series of specifications is subdivided into the functions and services shown in Table 5.4. A typical RF link budget for Ku band between a

Table 5.4
The DVB Family of Standards and Services

DVB-S (QPSK)	The satellite DTH system for use in the 11/12 GHz band, configurable to suit a wide range of transponder bandwidths and EIRPs
DVB-C (QAM)	The cable delivery system, compatible with DVB S and normally to be used with 8 MHz channels (e.g., consistent with the 625 line systems common in Europe, Africa, and Asia)
DVB-MS (OFDM)	The satellite master antenna TV (SMATV) system, adapted from the above standards to serve private cable and communities
DVB-T	The digital terrestrial TV system designed for 7–8 MHz channels
DVB-SI	The service information system for use by the DVB decoder to configure itself and to help the user navigate the DVB bit streams
DVB-TXT	The DVB fixed format teletext transport specification
DVB-CI	The DVB common interface for use in conditional access and other applications

high-power uplink earth station and a home-receiving terminal was provided in Tables 3.1, 3.2, and 3.3.

Since the topic of this book is satellite communication ground segments, the basic standard of most interest is DVB-S. It provides a range of solutions that are suitable for transponder bandwidths between 26 and 72 MHz, which includes all of the BSS and FSS satellite systems in existence or under development. It is composed of a layered architecture of which the payload contains the useful bit stream. As we move down the layers, additional supporting and redundancy bits are added to make the signal less sensitive to errors and to arrange the payload in a form suitable for broadcasting to individually owned, integrated receiver-decoders (IRDs). The system uses QPSK modulation and concatenated error protection based on a convolutional code and a shortened RS code. Compatibility with the MPEG-2 coded TV services, with a transmission structure synchronous with the packet multiplex, is provided. All service components are time division multiplexed on a single digital carrier. Bit rates and bandwidths can be adjusted to match the needs of the satellite link and transponder bandwidth, and can be changed during operation.

The DVB data stream is processed and assembled with the encoding system of the broadcast center. Video, audio, and other data are inserted into payload packets of fixed length according to the MPEG transport system packet specification. The structure is indicated in Figure 5.3 and reviewed next.

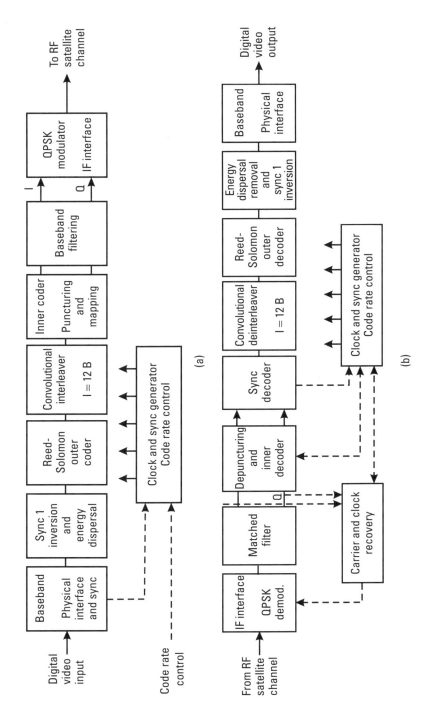

Figure 5.3 Functional block diagrams of the DVB-S (a) transmitter and (b) receiver.

- The MPEG payload is converted into the DVB-S structure within the multiplexer by inverting synchronization bytes in every eighth packet header (the header is at the front end of the payload). There are exactly 188 bytes in each payload packet, which includes program-specific information so that the standard MPEG-2 decoder can capture and decode the payload. This data contains picture and sound, along with synchronization data for the decoder to be able to re-create the source material.

- The contents are then randomized.

- The first stage of forward error correction (FEC) is introduced with the "outer code," which is in the form of the RS fixed length code. This transmits 204 output bits for every 188 input bits, representing approximately 8.5% overhead. RS is the most common outer code in use in this type of application.

- A process called "convolutional interleaving" is next applied, wherein the bits are rearranged in a manner to reduce the impact of a block of errors on the satellite link (due to short interruptions from interference or fading). Convolutional interleaving should not be confused with convolutional coding, which is added in the next step. Experience has shown that the combination of RS coding and convolutional interleaving produces a very robust signal in the typical Ku link environment.

- The inner FEC is then introduced by the modulator in the form of punctured convolutional code. In a punctured code, some of the bits of a normal convolutional "mother" code (e.g., $R = 1/2$) are deleted to reduce bandwidth (e.g., $R = 3/4$). Error correcting properties of a punctured code are nearly the same as an unpunctured code of the same rate. The amount of extra bits for the inner code is a design or operating variable so that the amount of error correction can be traded off against the increased bandwidth. This permits the coding rate, R, to be varied between $1/2$ and $7/8$ to meet the particular needs of the service provider. Specified values of R and the typical values of required E_b/N_0 are indicated in Table 5.5.

- The final step is the physical layer, where the bits are modulated on a QPSK carrier to be uplinked by the RFT.

The encoding process reduces burst errors through the randomization process. The amount of inner code FEC is adjusted for the link budget at the

Table 5.5
Coding Rate (R) and Typical Values of Minimum E_b/N_0 for Use in the DVB-S Standard
(The link is assumed to operate at 2×10^{-4} error rate)

Inner code rate	1/2	2/3	3/4	5/6	7/8
E_b/N_0, dB	3.3	3.8	4.3	4.8	5.2

particular frequency, satellite EIRP, transmission rate, and rainfall statistics for the service area. The standard allows the DTH operator to consider a range of availability requirements, including 99.7, 99.9, and 99.99%.

While DVB was established in Europe, the U.S. counterparts were standardizing how to digitize terrestrial broadcasting. Led by the FCC and a cooperative effort of U.S. broadcasters and electronics manufacturers called the Advanced Television Standards Committee (ATSC), the United States produced Digital TV (DTV) to incorporate both needs [3]. Satellite DTH services, on the other hand, have moved ahead on their own through the DIRECTV Satellite System (DSS) and, in the case of Echostar, DVB from Europe.

The ATSC defined the transport format and protocol for DTV to be compatible with MPEG-2. Based on a fixed packet length, it incorporates application (audio and video) encoding and decoding functions and the transmission subsystem. The basic video formats for standard definition and high definition TV are reviewed in Table 5.6. The specifics of the standard are beyond the scope of this book, since they involve considerable detail with respect to the frame format and syntax of the digital bit stream [4].

This review of digital TV requirements was brief, but it should give readers a flavor of what drives ground segment and earth station design for what is currently the highest-paying commercial satellite application. More detail can be found in the references. Also, an excellent source of current information is the conference and exposition of the National Association of Broadcasters, held annually in Las Vegas, Nevada, in April.

5.2 Data Broadcasting

Data broadcasting is a relatively small but potentially viable segment of the satellite industry, addressing specific needs for delivery of a common information format across a wide area. It plays to the strength of GEO satellites, particularly Ku band, since low-cost receive-only terminals may be deployed

Table 5.6
ATSC Digital TV (DTV) Video Format (analog input/output)

Total Vertical Lines	Horizontal Resolution	Aspect Ratio	Frame-Rate Coding, Hz	Line Sequence	Source of Standard
1,080 (1,080 lines actually encoded to satisfy MPEG-2)	1920	16:9, square pixels	23.976, 24, 29.97, 30	Progressive scan	SMPTE 274M
			29.97, 30	Interlaced scan	
720	1280	16:9, square pixels	23.976, 24, 29.97, 30, 59.95, 60	Progressive scan	SMPTE S17.392
480	704	4:3, 16:9	23.976, 24, 29.97, 30	Progressive scan	ITU-R Rec. 601-4
			29.97, 30	Interlaced scan	
	640	4:3, square pixels	23.976, 24, 29.97, 30	Progressive scan	
			29.97, 30	Interlaced scan	

and connected to equally low-cost PCs and other types of data devices. The explosion in the use of the Internet is playing a role in expanding the data broadcasting market, as exemplified by the DirecPC offering of Hughes Network Systems (HNS), Cyberstar of Loral Space and Communications, and the data capabilities of the DVB standard.

Data broadcasting applications are usually paid for on a subscription basis and fall into the following categories:

- Distribution of background music and advertising, which was originally accomplished on an analog basis but now can employ compressed digital audio and data. Pioneers like Musak and Supermarket Radio Network clearly demonstrated that one could use a satellite as an effective private broadcast station.

- Financial data (stock market "ticker" and commodity reports). This was originally an attractive application as it provided near instantaneous information to targeted audiences that could afford to pay for subscription.

- News wire services (e.g., Reuters and AP). These services are broadcast as sequential articles and other information, making satellite delivery nearly optimum.

- Data associated with DTH TV services. Examples include the online program guide, authorization for watching programming, and E-mail. Digital DTH systems are gearing up to offer the other types of data listed above as a means to increase their penetration and revenues. The fact that MPEG-2 and DVB are available widely is a factor in the success of this combination for bringing data broadcasting to wider markets.

- Internet access and Web page delivery over a high-speed broadcast link. As will be discussed, the DVB standard provides a satisfactory transport stream for IP broadcasting of content, overcoming limitations of terrestrial networks (e.g., the conventional Internet itself, with its thousands of routers).

Perhaps the most exciting domain of data broadcasting is that of the distribution of networked multimedia, the concept behind the World Wide Web. From the outset, the Web was envisaged as an open publishing network, where anyone could make their information (generally referred to as content) available to the public. According to François Fluckiger, one of the founders of the Web, multimedia can be defined as the field with the computer-controlled integration of text, graphics, still and moving images, and any other medium in which every piece of information can be represented, stored, transmitted, and processed digitally [5]. For the broadcast medium of satellite communications, multicasting means propagation of a flow of information from one source to only a subset of potential destinations. It is different from pure broadcasting in that it limits the reception, presumably through an addressing and control system.

5.2.1 Data Broadcasting with the DVB Standard

The European Telecommunications Standards Institute (ETSI) devised a specification for utilizing the MPEG-2 information transport to provide data broadcasting as an adjunct to television service [6]. It views data broadcasting as an important extension of its standard for possible applications like software download over satellite, cable, or terrestrial links, the delivery of Internet services over broadcast channels through IP tunneling, and interactive forms of TV. It identified the following data broadcasting profiles:

- *Data piping*—data pipes to support broadcast simple asynchronous end-to-end delivery of data through DVB-compliant broadcast networks (using the MPEG-2 TS packets).

- *Data streaming*—supports data broadcast services that require a stream-oriented end-to-end delivery in asynchronous (RS-232 interface) or synchronized (requiring a synchronize clock as in T1 or E1 transmission) mode.

- *Multiprotocol encapsulation*—supports transmissions of datagram communication protocols.

- *Data carousels*—supports data broadcast services that require the periodic transmission of data modules of known sizes that may be updated, added to, or removed from the data carousel at any time. This has been extended to object carousels using MPEG-4 and its object-oriented format.

5.2.2   Proprietary IP Satellite Multicasting

A practical implementation of data broadcasting for Internet access is the DirecPC service offering of HNS [7]. Available in North America, Western Europe, Japan, and other countries, DirecPC combines data broadcasting with the all-important return channel for controlling access and making requests for specific information transfer. As shown in Figure 5.4, the architecture relies on the satellite only to provide a wideband one-way broadcast to low-cost receivers, while the return communication path is by normal dial-up telephone via the Internet. This simple but powerful scheme provides Internet access much faster for downloads and comparable to what might be obtained from terrestrial high-speed access (HSA) using cable modems and digital subscriber line (DSL). In a test conducted by *PC World* magazine, the performance of various access technologies was measured with regard to download of a large image file [8]. Table 5.7 provides the results of this measurement.

The results in the table are not conclusive but do indicate that even a basic data broadcasting scheme can be competitive with a relatively wideband terrestrial connection of high quality. Actual performance will depend on numerous factors, including the dynamic loading of the satellite link or other shared bandwidth medium like the cable modem, the situation on the Internet and the originating server itself, and the performance of the protocol and application.

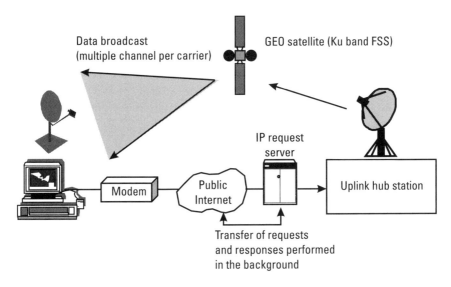

Figure 5.4 Architecture for IP data broadcasting and return channel via PSTN.

Table 5.7
Quantity of Data Received from an Internet File Download During a Nine-Second Period

Access Technology	Bulk Quantity of Data
Cable modem or T1	839K (the entire image file)
ADSL	629K
Satellite (DirecPC)	444K
ISDN (64 Kbps)	110K
56 Kbps (dial-up connection)	47K
33.6 Kbps (dial-up connection	31K

5.3 Digital Audio Broadcasting

First introduced in Africa in 1999 by Worldspace, Digital Audio Broadcasting (DAB) is a new satellite-delivered consumer service that provides the equivalent of multiple radio stations. Depending on the link budget, it can transport audio and other information in digital form to automobiles and other moving platforms. This is of potential interest to developing areas like central Africa, where the terrestrial infrastructure is limited, and to developed countries that could afford subscription broadcast services to vehicles on the move. Both of these applications were well into the implementation phase at the time of this writing but had not yet reached profitability from a business standpoint.

5.3.1 DAB Requirements

A satellite that provides DAB services would likely be of the area coverage variety, as illustrated for the XM Radio system in Figure 5.5. Operating at a downlink frequency of 2.333 GHz (e.g., S band), this satellite employs spectrum with reasonably good propagation characteristics for a mobile service. Since there will be situations where the signal is blocked by obstacles like trees and buildings, commercial service demands a scheme for providing a continuous flow of audio. This can be addressed in one of three ways.

- Prestore some quantity of data before starting to play. An internal buffer delivers a signal if the external signal is blocked, a principle also applied in CD players for automobiles.

- Provide two or more independent paths for the signal to reach the receiver (e.g., path diversity). Operating a pair of satellites that are sufficiently apart in the orbit provides considerable benefit.

- Install terrestrial repeaters (e.g., gap fillers) that receive the signal directly from the satellite and simply retransmit it locally. Since this is on the same frequency, it will introduce multipath signals into mobile receivers, which must have provision for selecting or constructively combining them.

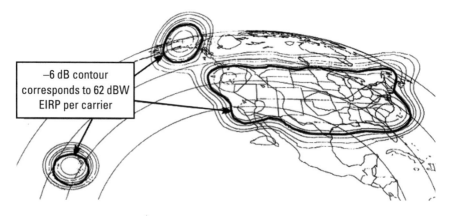

Figure 5.5 Coverage footprint of WorldSpace "XM Radio" DAR satellite at 115° west longitude (filing before the FCC).

The most likely schemes to be applied are the second and third because they provide much greater assurance of reception (and can be combined with the first, in any case). For the system with gap fillers, an implementation under study for Japan uses a form of CDMA to allow the receiver to recover information from multiple signals on the same frequency. By using two CDMA detectors, the rake receiver shown in Figure 5.6 captures two paths at the same time and can employ the valid data on either [9]. This name derives from the action of a garden rake, which collects leaves with multiple prongs.

5.3.2   Link Margin in DAB

A critical issue in the design of DAB systems is the available link margin to cover fading and shadowing. The amount will depend on the particular scheme used to maintain signal strength. In general, the more margin, the better; however, after a point, it becomes uneconomic to provide this excess power from the space segment. Research at the Jet Propulsion Laboratory has uncovered some effective ways to operate in this type of fading environment [10]. One of these uses a pair of receiving antennas on the vehicle, a technique called antenna diversity. As shown in Figure 5.7, two antennas mounted about two meters apart on the same vehicle can reduce the fade (in dB) by a factor of about two. For example, with a single antenna fade of 6 dB, a 2-m separation to a second antenna will cut the fade depth to 3 dB. This represents a saving of downlink power of the same amount, all other things being equal.

5.3.3   Vehicular Terminal Concepts

There is a requirement for receivers to be well integrated into the vehicle electronic system. In particular, the manufacturers of vehicles and their radios need to understand and support this new technology. General Motors, for example, introduced the OnStar system, which employs a permanently installed, hands-free cellular telephone and a GPS receiver to give the driver instantaneous help from a remote service consultant. In time, GM will upgrade OnStar for greater integration with the vehicle electronics, including the possibility of DAB reception. On this point, GM is lead investor in XM Radio, one of the two U.S. DAB operators.

The alternative is to provide some kind of attachment that is more or less independent of the vehicle electrical system. Anyone who has tried to use a portable CD player in a car knows that this is less than optimum. One

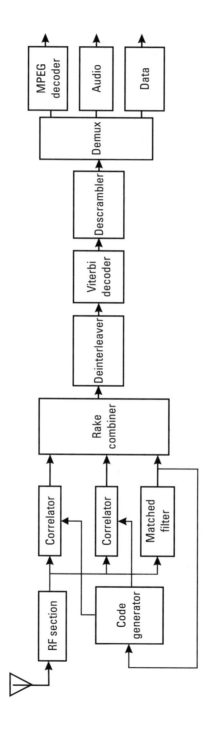

Figure 5.6 Block diagram of a rake receiver, which extracts useful portions of multipath signals at particular time delays.

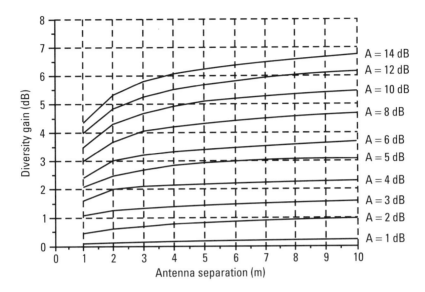

Figure 5.7 Diversity gain versus antenna separation distance for a family of single antenna fade levels, based on rural roadside tree measurement campaigns [10].

measure of success for DAB will be its ability to become part of the vehicular electronic systems so that the audio and data can flow efficiently and be provided to users in a friendly and welcome manner.

References

[1]   http://www.it.ed.tudelft.nl/itcolleges/et10-38/hc/hc13/sld057.htm

[2]   http://www/dvb.org/dvb_framer.htm

[3]   Whitaker, Jerry, *DTV: The Revolution in Electronic Imaging*, Video/Audio Profession Series, New York: McGraw-Hill, 1998.

[4]   Proceedings of the 1998 Broadcast Engineering Conference (including the ATSC standard for terrestrial broadcast and cable), April 5–9, Las Vegas, NV, National Association of Broadcasters, Washington, D.C., 1998.

[5]   Fluckiger, François, *Understanding Networked Multimedia: Applications and Technology*, London: Prentice Hall, 1995, p. 589.

[6]   European Telecommunications Standards Institute, "Digital Video Broadcasting (DVB); DVB Specification for Data Broadcasting," TS 101 192 V1.1.1 (1997-10).

[7]   Dillion, Douglas, "Asymmetrical Internet Access over Satellite-Terrestrial Networks," AIAA 16th International Communications Satellite Systems Conference, a collection of technical papers, Part 1, Washington, D.C., American Institute of Aeronautics and Astronautics, Washington, D.C., 1995, p. 476.

[8]   Spanbauer, Scott, "Bandwidth on Demand: The Fastest Ways to Connect, from 56-Kbps Modems to Satellite Hookups," *PC World*, August 1997, p. 159.

[9]   Kikuchi, Hideo, Yoichi Koishi, and Yukiyoshi Fugimore, "Digital DBS System for New Multimedia Era," AIAA 17th International Communications Satellite Systems Conference, a collection of technical papers, Yokohama, Japan, American Institute of Aeronautics and Astronautics, Washington, D.C., 1995, p. 525.

[10]   Goldhirsh, J., and W. J. Vogel, *Propagation Effects for Land Mobile Satellite Systems: Overview of Experiments and Modeling Results,* Washington, D.C.: National Aeronautics and Space Administration, Office of Management, Scientific and Technical Information Program, 1992.

6

Ground Segment Baseband Architecture

The ground segment architecture of a communication satellite blends telecommunications engineering principles with the tools and techniques familiar to satellite engineers. Therefore, one needs to understand how these fields are applied during all phases that define, specify, implement, operate, and maintain these networks. As covered in Chapters 4 and 5, ground segments vary in function and complexity, reflecting the services to be offered and the details of implementation. Here, we consider common issues concerning the architecture of the ground segment before moving to the detailed design of earth stations and user terminals. We generally find that the basic functionality of architecture is embodied in the digital processing and baseband elements, which are reviewed in this chapter as well.

The main elements of ground segment architecture are illustrated in Figure 6.1, indicating the user, gateway, and management components. To provide telecommunication services, the network allows users to interact with one another by talking; exchanging E-mail, fax, or image data; or interacting with data transfer associated with the Internet or a video teleconferencing system. The latter represent evolving approaches to telecommunication services and might someday replace pure voice service common in the PSTN. Broadcast (point-to-multipoint) connectivity, the principal alternative, offers the means to deliver information content to the wide audience defined by the satellite coverage. The architectures for the two ground segment roles are usually different, taking account of special features to be described in this chapter. Still, Figure 6.1 suggests that for typical digital ground segments which directly serve users, there are common elements that one expects to find almost universally:

- *A central facility* to broadcast information or provide access to the terrestrial networks within a country or region. The former might be self-contained in the sense that the information is either created or replayed locally. More likely, this facility must itself connect to other sources of content, either through backhaul links provided by terrestrial networks or other satellite networks. The size and extent of the central facility grows as broadband services begin to dominate the application mix.

- *User terminals* of appropriate size and complexity (and cost) to satisfy the local needs of subscribers or other classes of users. These might be owned and operated by individuals or provided to remote branches for use in a corporate or government network. There is a trend to make these terminals cheap and easy to operate so that they can be placed at every location in which services are required. On

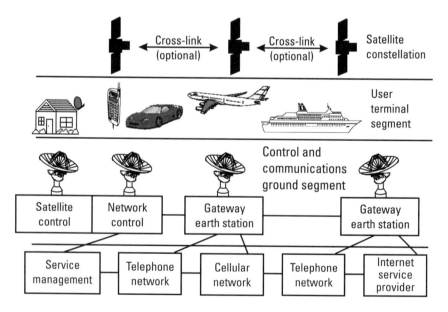

Figure 6.1 Main architectural elements of a communications ground segment (fixed or mobile).

the other hand, there will be circumstances where user terminals must be shared to reduce investment and operating costs. Such terminals may begin to take on the properties of larger earth stations of more traditional design.

- *A network control facility* to coordinate the provision of services to users and manage the overall system on a day-to-day basis. A network control facility could be collocated with a hub earth station or broadcast center (as appropriate), or might be an independent station that can access the space segment directly. Design of the network control and management function is perhaps the most complex aspect of the ground segment, since it is highly interactive with the information and users. This may involve direct monitoring of space segment RF signals from bent-pipe transponders (i.e., transponder management) or focus exclusively on baseband information processing and flow, as in the case of telephone traffic or IP data management.

6.1 Architecture Definition and Design

System architecture definition and design encompasses a number of disciplines that have been defined in previous chapters. In this section, we go into more detail concerning the most critical of these approaches, as they can determine the acceptability of the ground segment once users gain access to it. Included are the following principles:

- A clear definition and flowdown of service objectives, in terms of quality and quantity;

- Adequate determination and allocation of signal impairments that affect many of the most highly leveraged performance characteristics of the ground segment;

- Identification of the topology and distribution of the baseband elements of the system;

- Creation of a service management approach, using the hardware and software aspects of the architecture, adapted to the organization's needs and abilities.

6.1.1 Meeting the Service Objectives

The requirements of the ground segment have been expressed in terms of user services that can be accessed either directly from the space segment (by user terminals on the customer's premises or directly held) or through shared earth stations (e.g., hubs, gateways, and the like). Systems architects and their business counterparts in marketing and finance attempt to define what service mix would be of value to potential subscribers. They usually have to make many assumptions if a real system (satellite or terrestrial) does not already exist.

Selection of appropriate market segments, which are both characteristics and behaviors, is both a science and an art [1]. By characteristics, we mean such things as age, sex, income, lifestyle, occupation, and location of residence and nationality. By behaviors, we mean appeal based on benefits (prestige or convenience of use), experience of user, amount of usage, and attitude toward the product or supplier. For example, the first proponents of global mobile personal communication service (GMPCS) via satellite viewed the international business traveler (IBT) as the most likely and lucrative target. The characteristics of such potential users were researched over the years and appropriate properties ascribed to them. Examples include: the number of trips per year, where they traveled (to developed and developing regions of the world), their dependence on cellular telephone services while away from their offices, and how much they would be willing to pay for seamless service. This type of information feeds back to indicate the number of subscribers, where they would use their GMPCS phones, and the aggregate calling (and revenue) that one might reasonably assume.

In time, the GMPCS operators came to realize that there probably would not be enough IBTs to go around (the number was further reduced by the rapid adoption of GSM in the more heavily traveled regions of Europe, Asia, and North America). The operators discovered that other uses could be significant and potentially of greater value in the aggregate. This produced the type of market segmentation shown in Figure 6.2. Another important aspect is the rollout of service and the resulting ramp-up of revenues as one or more of the systems are adopted by various market segments. This is suggested in Figure 6.3 for a hypothetical system with global coverage. If the service is targeted to a single country or region, then the corresponding behavior might look the same in principle but could be very different in specific terms. The variability of these important parameters will determine if a given ground segment reaches its financial breakeven point and eventually makes money.

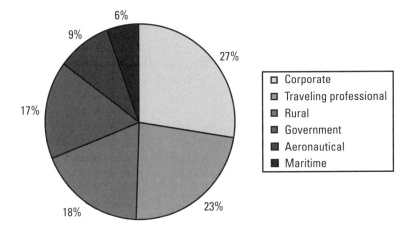

Figure 6.2 Conceptual target market segments and their estimated relative size in terms of subscriber count (for GMPCS system).

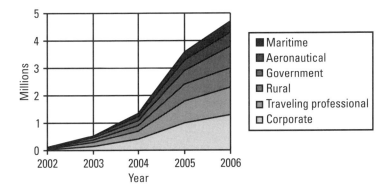

Figure 6.3 Hypothetical ramp-up in subscriber count for target market segments shown in Figure 6.2.

Analysts can create the spreadsheets and graphs, but the numbers they use should be based on reality or at least valid research. The former has been the domain of established telecommunication operators, including satellite operators like INTELSAT and Inmarsat, whose experience provides confidence in the planning process. Once competition is introduced, however, even established systems and ground segments cannot be assured by assumed financial models. This is why it is always necessary to look at many scenarios and to perform sensitivity studies to uncover the critical parameters that affect the desired outcome. For the case of a new concept without a valid precursor, the study becomes much more of a gamble.

There is a value to market research in identifying and measuring the potential attractiveness (or unattractiveness) of a new service offering. Surveys, focus groups, and pilot trials can be devised and perfected as means of producing comparative and quantitative information to assess the value of new services to potential subscribers. If done in an artful way, market research can be combined with engineering techniques such as the mean opinion score (MOS), which was discussed in Chapter 4. Statistical techniques exist that allow the systems analyst to find the optimum bundle of services, considering service features and their cost to potential users. A popular technique called *conjoint analysis* links features directly to preferences or to customer needs [2]. The survey allows potential buyers to rate the relative value or utility of each feature (including what they are willing to pay). For example, for a hypothetical telecommunication service that provides image transmission, the profile of a particular package of features is valued as

$$P_m = u_r + u_a + u_h + u_t + u_b$$

where u_r = part worth of resolution of fax output; u_a = part worth of service accessibility to user; u_h = part worth of availability of hard copy output; u_t = part worth of transmission time (the shorter, the higher the score); and u_b = the base level of preference for the image service.

The part worth values are determined through the survey of a sufficiently large population, where potential users are exposed to or asked about each of the features. They rank them according to their potential interest in obtaining the service for either business or personal use. Once the utilities are known, the attractiveness of a given product combination is found by summing the values. The package with the highest sum will probably be the most attractive to the targeted user population.

All of these methods help the system analyst or developer examine a potentially important aspect of the service. However, they cannot provide guarantees of what will succeed in the marketplace. This is further blurred by the fact that market segments change over time due to user preferences and, of course, competition from potentially better services. These factors can be taken into account by designing flexibility into the ground segment architecture, to allow changes in service type and mix and to provide for growth.

Once the system developer has clarified the market segments and applications to be delivered, the next challenge would be to size the entire ground segment. This includes the network as a whole and any major earth stations that are points of traffic aggregation. Individual user terminals might be sized for the smallest division of service and so would be considered in aggregate as far as how they would load the space segment and major earth stations. The network control and management system would then have to support this overall requirement as well. The following two sections address the two key aspects of this sizing issue.

6.1.2 Quality of Service (QoS) in Broadband Networks

Once you know what services are to be provided and where, the next most critical set of parameters deals with the quality of service (QoS) as experienced by subscribers or other users that receive service from time to time. While this may appear to be a technical matter, it becomes deeply involved with human nature, preferences (as discussed above), and comparisons with whatever else is out there to experience. For telephone networks, QoS can be measured simply in terms of the following characteristics (reviewed in Chapter 4):

- Voice quality, as measured using mean opinion score (MOS);
- Call blocking probability;
- Post dialing delay;
- Availability.

All of these are measurable and relatively repeatable. The problem becomes more complicated when considering a service with more features than POTS (plain old telephone service). For example, with cellular and PCS services, one is also concerned with the ability of subscribers to maintain a connection as they travel around a cell and as their call is handed off between cells. Cellular network engineering cannot guarantee 100% maintenance of the connection in this dynamic environment, due to the variability of radio coverage and performance of the mobile user terminal. The problem becomes even more complex when considering GMPCS, digital DTH TV, Internet/WWW, and interactive multimedia. The QoS for a particular application that the user is engaged in will differ from another (e.g., another application for the same user, and even the same application for another user). This demands a dynamic network that allows the user to obtain an appropriate QoS.

While this is currently an ideal, it nevertheless deserves the attention of developers of broadband ground segments. François Fluckiger, the noted author on networked multimedia systems, offers the following general definition of QoS [3]:

> Quality of service is a concept based on the statement that all applications need the same performance from the network over which they run. Thus, applications may indicate their specific requirements to the network, before they start transmitting information data.

Another more practical definition is that used as a basis for B-ISDN, as being implemented in Europe:

> QoS is the measure of how good a service is, as presented to the user. It is expressed in user understandable language and manifests itself in a number of parameters, all of which have either subjective or objective values.

From a system design standpoint, QoS couples into some key performance parameters, some that are hard and some that are somewhat soft. Hard, or compulsory, requirements include:

- User performance guarantees, such as synchronous timing; end-to-end maximum delay, receipt of message; and errored bits, packets, or messages;

- Network operation properties, such as routing of packets to avoid congestion and ability to accept or reject a request for connection due to network state;

- Statistical properties, such as average and standard deviation of connect time, holding time, and call continuity.

Obviously, there are as many different QoS measures as there are service characteristics, and the number multiplies as one considers the range of possible applications and user profiles. This is an area of continuous study in broadband network standards, with much of the more interesting work being performed in the context of ATM and B-ISDN, including criteria such as:

- Cell transfer delay, which is the end-to-end delay through the network nodes and links;

- Cell delay variation, the statistical nature of cell transfer time due to its asynchronous nature and the variability of traffic loading;

- Cell loss ratio, caused by errors introduced in the network or cells dropped due to network congestion;

- Mean cell transfer delay, the arithmetic average of a specific number of cell transfer delays;

- Cell error rate, which is the ratio of errored cells divided by the total cells (e.g., successfully transferred cells and the errored cells);

- Severely errored cell block ratio, which measures errors in blocks of cells and is the ratio of the total severely errored cell blocks divided by the total transmitted cell blocks;

- Cell misinsertion rate, which is the ratio of misinserted cells to the time interval (a misinserted cell is a received cell that has no corresponding transmitted cell on that particular connection).

The QoS definitions for ATM are very technical and specific to the nature of this protocol. They may have little direct relevance to users who are interested in top-level performance, as discussed previously in this section. However, it is the duty of the system architect to make the right connections between these two worlds of QoS. A ground segment for which this is done correctly will be more likely to succeed in both technical and financial terms.

6.1.3 Capacity and Distribution of Services

The market segments, services, and QoS requirements having been set, the key remaining attributes of the system to be defined are the capacity of the ground segment and the corresponding distribution of this capacity to the various locations within the overall network. This will determine the economics of the service when combined with other financial matters such as investment, operating costs, cost of money, and pricing. Selection of locations for service delivery is a very complex matter, being determined by such factors as:

- Cities, regions, or countries where the service provider has gained market access;

- Regulatory requirements of the location or region, including availability of spectrum and procedures/costs for obtaining it;
- The local need for services of this type, reflecting the tastes and specialized needs of local subscribers and businesses;
- The ability of potential users to pay for equipment and services;
- Competition from local service providers who offer comparable services;
- Competition from other satellite-based service providers;
- The practical aspects of implementing a suitable ground segment when and where needed;
- The ability of the ground segment operator to manage the assets and service delivery.

The design of the space segment in terms of the orbit configuration, antenna and repeater arrangement, and communications processing has a large impact on how capacity and services are allocated across the service area. For example, the ability of a single GEO satellite to perform these functions is determined by its overall capability in terms of RF radiated power (EIRP) across the coverage region and the quantity and mass of payload hardware that are employed to relay signals of the ground segment. For the spacecraft antenna system, the coverage footprint is key to defining the potential of the satellite to support the service and its distribution. It is these aspects that must be carefully considered in the design, because changes in orbit are typically very limited in nature. In time, satellites will be designed with greater flexibility for moving capacity and services around the visible globe. This is the motivation for satellites that employ digital processing repeaters, phased array antennas, and quantities of spot beams that can be arranged to suit the dynamic distribution of traffic.

Non-GEO satellite constellations offer special promise of making the space and ground segments almost independent of each other with regard to location. With satellites streaming past almost every corner of the globe, the earth stations and user terminals can be placed for optimum service to local regions and populations. A constellation like Iridium with intersatellite links can simplify service provision and distribution, since any satellite allows a user terminal or gateway to access the global infrastructure. LEO bent-pipe systems without such links, like Globalstar and Skybridge, are simpler to construct yet impose the restriction that the user terminal and associated gateway earth station must be in view of a common satellite. Of course, this satellite would change during the normal course of a connection, requiring some form of handoff. One then needs a gateway in the same country or region, which results in patterns of national service (which allows the features to be further customized to match the local market). The ICO system uses MEO and therefore permits regional services to be delivered from a single satellite, much the way one expects from a GEO satellite.

There is a clear trend in space segment design to employ multiple spot beams to achieve a number of aims:

- *Reuse frequency to increase satellite and space segment capacity.* This reduces cost per unit of service and allows the satellite to meet greater total demand.

- *Increase the EIRP and G/T of the satellite to reduce the cost and size of earth stations and user terminals.* This is what has allowed GMPCS systems to provide services to handheld mobile satellite phones. In a similar manner, Ka-band broadband systems will introduce user terminals with compact antennas and low power transmitters that deliver high-speed Internet services and other innovations like personal video conferencing.

- *Provide coverage of major cities and other "hot spots" of demand.*

A difficulty that can exist is that the capacity of a given beam is determined by the available bandwidth to the beam and the characteristics of the air interface (e.g., the link budget, modulation, and multiple access method). Once the capacity of one beam is reached, it is not possible to uplink and/or downlink traffic to or from that particular beam. Due to the constraints of the beam-to-beam routing patterns, saturation of such a highly demanded beam can limit the aggregate capacity of the satellite and network. Means to overcome this limitation include: using more bandwidth-efficient modulation or multiple access, adding more spectrum (either in the same but adjacent band or in a completely different band), or adding satellites in such a way as to not introduce unacceptable interference into the operation of the existing satellite(s). None of these solutions are particularly easy, but they may become necessary as traffic builds.

While the satellite determines the overall coverage area, the specific installation of major earth stations usually defines how services might be delivered and where. Tied together with the satellite, the ground segment represents a fixed base-station infrastructure to allow user terminals to access the network. Unless one is speaking of mobile or transportable earth stations, the ground segment will represent a sunk cost that cannot easily be redeployed to different locations or applications. An approach to overcome this problem is to install major earth station equipment in an existing facility such as a teleport. User terminals, on the other hand, are designed more like consumer electronic items and can be discarded when the service is not desired or available.

6.2　Baseband Architecture

The functionality of the ground segment architecture is principally embodied in the baseband equipment contained within the earth stations and user terminal. Reflecting this point is the fact that suppliers of application systems for VSATs, DVB services, IP multicasting, and fixed telephony have concentrated on the baseband sections, leaving the RF terminals to be supplied by other manufacturers and integrators who are familiar with the microwave industry. In Table 6.1, we provide a general outline and definition of the functions performed by the baseband architecture. In coming years, some of these functions will be allocated to a digital processing space segment, as is the case for Iridium. For bent-pipe satellites, all baseband functions are, of necessity, performed on the ground. Figure 6.4 provides a generalized baseband architecture, based on the previous assumptions, and the following

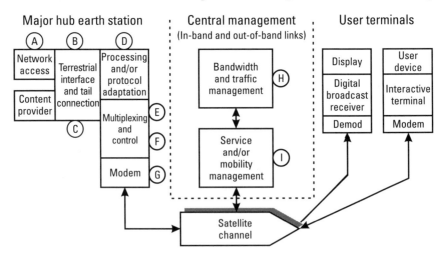

Figure 6.4　Generic baseband architecture for a digital communications satellite network.

Table 6.1
Allocation of Requirements and Functions to Baseband Architecture and Earth Station Elements

Function	Requirement	Element	Description
User interface or access to content	Deliver a standard application to the user (or access content)	A	Defined by selected standard for the particular application
Backhaul circuits	Connect between user interface (or content source) and earth station	B	Typically a terrestrial transmission line using cable, fiber, or local microwave
Terrestrial interface	Interface from backhaul circuit to the baseband equipment of the earth station	C	Port cards or connectors that comply with the selected standard (in A)
Information processing	Compression, encryption, protocol conversion, etc., based on unique constraints of the baseband architecture	D	Data processing or specialized protocol conversion equipment
Multiplexing	Combining two or more information channels into a single data stream	E	Multiplex equipment—fixed time slot, statistical or packet multiplex
Multiple access processing and control	Assign and allocate data for efficient transfer via the satellite network	F	Baseband multiple access equipment
Modulation and demodulation	Encode and transfer data streams between the baseband equipment and the RF terminal	G	Modem equipment
Bandwidth management	Engineer and optimize the flow of network traffic	H	Network control elements
Service management	Coordinate application delivery, information supply, and conditional access for users	I	Service management elements

paragraphs review some of the important issues involved in the definition and selection of the associated nine elements.

6.2.1   User Interface or Access to Content

The first point of entry into the ground segment architecture is the specific interface at the user (voice, data, videoconferencing, etc.) or, in the case of broadcast applications, the source of content (TV programs, Web pages, radio programs, etc.). We previously reviewed some of the critical demands on the user side in terms of QoS and the like; here our concern is the electrical and procedural aspects ensuring that the service will connect properly. The user interface will generally satisfy a recognized standard (such as RS-232C for low- or medium-speed data, RJ-11 modular jack for analog telephone, V.35 for medium- or high-speed data, and Ethernet for LANs), consistent with low or moderately priced terminal equipment and terrestrial telecommunications networks around the world. The data format, protocol, and signal characteristics at this interface will adhere to a standard format that is compatible with the user device.

6.2.2 Backhaul Circuits

The basic function of a backhaul circuit is to provide the "last mile" link between the earth station and the actual user or source of content. These circuits can be a few kilometers in length or perhaps thousands of kilometers for extensive terrestrial arrangements. If user and earth station are co-located, the backhaul circuit is replaced by a piece of cable of appropriate length. Backhaul circuits may be owned, shared, leased, or obtained as a service from a terrestrial service provider (particularly telephone companies). The key is to obtain sufficient bandwidth with the needed quality and reliability at an acceptable price. The latter must consider the investment cost as well as the annual operating costs.

Options for backhaul circuits are derived from what is available on the market or can be constructed in reasonable time. These include the following:

- Twisted-pair cable;
- Coaxial cable;
- Fiber-optic cable (multimode);
- Fiber-optic cable (single mode);
- Line-of-site microwave (point-to-point);
- Local microwave distribution service (LMDS, point-to-multipoint).

Twisted pair, coaxial cable, and multimode fiber are normally associated with short distances (e.g., under about 2 km) within a building or campus environment. Telephone companies will use twisted-pair and coaxial cable for the last mile as part of a long-haul connection. Once the signals are inside their plant, they would normally multiplex the data onto single-mode fiber-optic cables to take advantage of the existing investment and the favorable per-circuit economics offered thereby. Microwave links are still applied for short-haul local connections for data rates up to about 1 Gbps; however, rarely does one find long-haul microwave in use today. The MMDS and newer LMDS technologies could be attractive in the short-haul case if a network is already up and in operation in the local area. Not to be overlooked is the ease, or lack thereof, of monitoring and managing this part of the architecture, as the service cannot be provided in the absence of a working backhaul.

The challenge for the ground segment architect is to first identify the backhaul requirements in terms of the locations to be connected and the bandwidth requirements. From there, it is possible to compare the viable alternatives as cited above, the providers of the services and/or equipment, and the costs of implementing and operating this segment of the system.

6.2.3 Terrestrial Interface

The end of the backhaul circuit that terminates at the earth station is called the terrestrial interface. As with the interface at the user or content provider, this interface would likely satisfy a recognized standard (the same standard, in fact). It is the function of the earth station baseband equipment to work within the constraints of the terrestrial interface and deliver a satisfactory service. In this way, the terrestrial interface shields the backhaul circuit and end user from the peculiarities of satellite communications in general and the implementation of the overall network in particular.

There are two options for how the terrestrial interface could be implemented at the earth station. In the standard arrangement, the interface is provided on a conventional termination frame such as is found in a telephone exchange. In the case of digital services, the interface is at a patch field. The alternative is for the connection to be made directly on the baseband equipment at what are called port cards. These are typically computer-type printed circuit cards with standard connectors on the rear edges. The backhaul circuit connection is made directly on the card, reducing the amount of rack space and cabling within the earth station. A potential drawback of this approach is that the backhaul circuit possibly cannot be tested or reconnected without disrupting the operation of the baseband equipment. In the case of a fiber backhaul circuit, it is likely that standard multiplexing equipment would be introduced at the terrestrial interface to separate individual digital channels (this is not to be confused with the multiplexing function contained within the baseband equipment). Other issues concerning lightning and surge protection for the backhaul circuit are addressed in Chapter 10.

6.2.4   Information Preprocessing for the Baseband Equipment

Information processing in the context of the baseband equipment is often the most compute-intensive aspect of the architecture. What we find within this element are complex functions like digital compression to reduce bandwidth and encryption to achieve security and control delivery of service. Protocol conversion is another important function for data networks such as those that support IP transfer. The techniques for performing this conversion depend heavily on the architecture of the network as well as other specifics that are up to the supplier of the hardware and software. A common technique is that of protocol spoofing, where the user and terrestrial interface are made to believe that they are directly connected to the other end of the satellite link when, in actuality, their digital conversation is with the information processing function at the closely connected earth station.

The type of equipment in use for information processing often consists of specialized digital electronics that provide cross-connections along with a high-speed bus and input/output functions. There would be a local processor with custom software to direct these functions, some of which could be performed within the processor itself. In many respects, the capability described here is much like that of an IP router of the type used in the Internet. The difference, however, is the way the equipment is configured and the software that runs it. Behind this capability there may be a server running a standard operating system such as UNIX or Windows NT, performing the overall management of this portion or all of the baseband equipment. The software in the server will, from time to time, be upgraded to take account of bug fixes, improvements, and new features to be included. The server will often contain the traffic routing map that organizes and routes information to the uplink and between locations.

6.2.5   Multiplexing or Packet Routing

The multiplexing function takes the required number of processed information streams and combines them to produce a high-speed baseband. In broadcasting, this stream represents several MPEG-2 channels along with associated data and control channels. The equivalent function in a telephone gateway using multiple-channel-per-carrier (MCPC) transmission would be to combine several voice channels using time division multiplex (TDM). In a variant called statistical multiplex (STATMUX), data from the input channels is only sent when the particular channel is active. This reduces the output data rate to some fraction of what would have been required (e.g., less than the sum of the input data rates). On the receive end, the demultiplexing process re-creates the independent streams from the buffered input. Some loss of information will occur if the STATMUX drops any bits when the sum of the inputs exceeds the maximum allowed.

Packet switching is gaining popularity over the STATMUX based on the greater availability of ATM technology. The scheme operates much the same as STATMUX in that the input channels are only connected through when they are active. The refinement here is that the input data is first packetized using ATM or another (possibly proprietary) packet protocol. The multiplexer in this scheme is operating as a packet switch or router, taking active packets from the information processor and applying them to the uplink as they arrive. The behavior of this arrangement is like that of an IP router and, in fact, router technology is directly applicable. However, the baseband architecture requirements for software may result in a different design approach.

6.2.6 Multiple Access Processing and Control

The portion of baseband architecture most critical to satellite network operation deals with multiple access (MA) control of transmission. Because of the unique properties of FDMA, TDMA, or CDMA and, more importantly, the particular implementation, the MA function integrates with much of the rest of the baseband system. When using TDMA, the functions of information processing and multiplexing are coordinated in such a way that transmissions from different earth stations do not interfere with one another when they reach the satellite. As discussed in Chapter 4, TDMA places stringent requirements on network timing to prevent signal overlap and maximize throughput. This is further complicated if the satellite contains multiple beams and possibly a time-division satellite-switch to interconnect them. From a logical standpoint, the information is assigned to time slots, which are routed through the shared bandwidth of the satellite repeater. Individual stations employ specific time slots for point-to-point traffic; they must adhere to a common network connectivity map to assure a consistent distribution of time slots and capacity. From time to time, the map is changed (manually or dynamically) to accommodate changes in the station and service mix.

When looking at FDMA, we find a much simpler type of system from the standpoint of MA control. In fact, the multiplexed information may be connected directly to the modem and uplink without any active control. This is possible because FDMA uses separate, noninterfering frequencies that can be assigned on a permanent or at least scheduled basis. The complexity and need for control comes with demand assignment (DA) of carrier frequencies under some type of traffic control. When operating in this manner, FDMA shares some complexity with TDMA.

CDMA is deceptively simple from a control standpoint, since individual transmissions use separate codes and hence do not directly interfere with each other. The need for control increases with the number of simultaneous transmissions (users). The objective of the architecture is to maximize the system capacity (i.e., the number of simultaneous users), which in turn requires that the transmitter be adjusted so that the signal-to-interference ratio at the satellite repeater is at a minimal acceptable level [4]. Ideally, earth stations and user terminals should maintain this desired level, otherwise capacity will be reduced (because there will be more interference noise than budgeted). This ideal is not possible in practice because of normal errors in aligning uplink power and the presence of uplink fades due to a variety of sources (particularly shadowing in mobile satellite communications). As a result, an effective uplink power control scheme must be introduced. Another feature now implemented in Globalstar is to adjust the coding and compression on the individual channels in response to the overall loading of the transponder. These factors are very complicated and comparisons with TDMA are both difficult and often inconclusive.

6.2.7   Modulation and Demodulation

The term *modem* has come to refer to the combination, in one package, of the modulator and demodulator that provides the conversion from bidirectional data to the send-receive transmission on a carrier frequency. In ground segment architecture, the actual modem may consist of separate modulator and demodulator due to the special attributes of the MA system and space link. For example, in the case of a conventional VSAT star network, the hub transmits a continuous TDM stream on the uplink, while the inroutes transmitted by the remote VSATs are in TDMA burst format at a significantly lower data rate (typically a ratio of 4:1 of outroute to inroute). In a given network supported by one outroute, there would be one outroute modulator transmitting a continuous TDM data stream, and multiple inroute demodulators each operating in a burst mode to recover the TDMA transmissions from multiple VSATs on the particular carrier frequency.

In some forms of FDMA, a combined modem can be used if the uplink and downlink bandwidths are the same. Carrier transmission is continuous, which is also more conducive to containing the send and receive functions within the same unit. The situation comes apart, however, in FDMA networks using what are termed multidestination carriers (MDCs). As practiced in the INTELSAT system using analog frequency division multiplex (FDM), a given MDC from one earth station contains groups of voice channels destined for several different destination earth stations. Since the carrier can be downlinked by a wide beam over the earth, any station in communication with the uplinking station would receive the carrier and recover only those channels intended for it. With N stations in the network, there would only be N carriers uplinked (instead of $N(N - 1)/2$, which would be required to produce a full mesh). However, a given station requires one modulator along with up to $N - 1$ demodulators, the precise number corresponding to the number of other stations with which it is communicating.

This brief discussion reviewed the architectural aspects of modulation and demodulation. Later in this chapter, we review the specific requirements and design approaches for modems as used in satellite communications. An important function often consigned to the modem is that of error correction in the form of convolutional encoding and decoding.

6.2.8 Bandwidth Management

The remaining functions of architecture relate to the transfer information from end to end and the management of the key resources of the system. Bandwidth management contains a number of system management functions for the network or broadcast operation, including traffic control, capacity assignment and optimization, and coordination between operating elements. The TDMA network connectivity map mentioned in Section 6.2.6 would be generated and maintained as part of bandwidth management. Depending on the organizational/political structure of the network, bandwidth management could be centralized (which is usually most efficient) or distributed among the stations (a less efficient—but possibly more reliable—alternative).

Bandwidth management in FDMA and CDMA involves assigning individual channels as needed to meet traffic demands. The requirements could be for fixed bandwidths that remain in place infinitely, in which case the bandwidth management function is satisfied with basic carrier monitoring. For demand-assigned applications, bandwidth management is a dynamic process similar to maintenance of the network connectivity map in TDMA.

6.2.9   Service Management

Service management is a vital function of the architecture and often it is treated more like an afterthought than a serious requirement. However, after the network is working and providing services, the operators quickly learn how effective their service management capability really is. If the requirements were set correctly in the beginning and the capabilities were incorporated consistently and strongly, then the provision of services will occur as needed. Under service management we find the important technical role of network monitor and control (to identify, troubleshoot, and resolve problems). The complete set of service management functions is potentially extensive, tied to the nature of the business structure, regulatory environment, and competition.

Our discussion of baseband architecture was from an overview perspective rather than going into particular approaches, which are usually proprietary. This is because of the extensive detail that is involved in the elements and interconnections needed to produce a working architecture. What is most appropriate for the architect is to be sure of the business and service requirements, the functions to be performed and their locations (geographically and logically), and the high-level solutions that can be applied (e.g., the MA system, baseband structure and protocols, and vendor technology platforms).

6.3   Baseband and Modem Equipment

The portion of the earth station that exists between the RF terminal and the terrestrial interface is defined as the baseband and modem equipment. A simplified block diagram of these elements is provided in Figure 6.5, indicating the modulator and demodulator (which could be combined in the same modem unit, depending on the application); the multiplex and demultiplex equipment; and a segment that we refer to as multiple access and timing control. Connected to the latter is the terrestrial interface that would apply if this earth station were to connect to an external network. Alternatively, the equipment may deliver services directly to a user device such as a TV or PC.

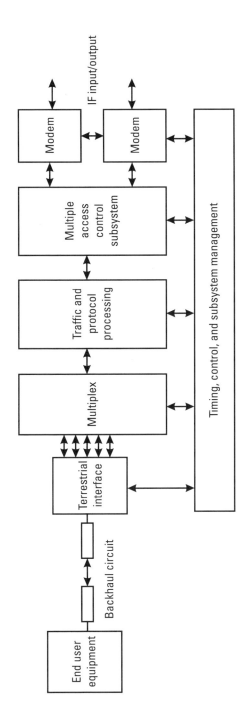

Figure 6.5 Baseband subsystem of a central hub earth station for digital communications services.

6.3.1   Baseband Equipment Design

Engineering principles for this important segment cover many fields, including analog modem design, digital signal processing, digital hardware design and development, and software engineering. Unlike RF engineering, which is driven by the requirements of the microwave link requirements, baseband equipment is highly integrated with the applications that the earth station supports. This means that this portion of the earth station is customized for the network and use. Our main consideration in this section is with the requirements for and performance of baseband equipment as compared to the RF terminal and satellite link. The engineering principles behind digital baseband equipment have developed from telecommunications switching along with strong input from the computer industry. Any baseband system is primarily embodied in an information processing structure that supports the collection, routing, management, and control of data. Underlying all of this is a strong element of software, both in terms of computer code and firmware that operates a wide variety of locally connected and remote devices throughout the baseband architecture.

Critical steps in baseband design include the following:

- Clear definition of the service requirements, operating modes, failsafe processes, and management aspects of the baseband architecture taken to the equipment level;

- Allocation of functions to components within the baseband equipment for earth stations and user terminals;

- Careful specification of component technical requirements, operating conditions, and the tradeoff between hardware and software implementation;

- Arrangement of the components within the baseband equipment and the interconnection/intercommunication processes;

- Selection of the optimum hardware and software platforms, considering make versus buy issues (i.e., build internally versus purchase of off-the-shelf solutions);

- Proper definition of timing and sequencing of functions;

- Design and functional verification of each component, considering also the operation and maintenance as a unit and as part of the baseband equipment;

- Creation of an operating control system and supporting man/machine interface for management of the baseband equipment and its functionality.

These tasks can represent a major effort and probably cannot be carried out in a vacuum. In particular, this will be a multidisciplinary effort involving systems architects, service designers, equipment designers, software engineers, and operations and maintenance specialists. Many of the tasks must be performed in parallel with others and considerable iteration will be needed. Often, a proof-of-concept pilot must be developed, possibly involving a hardware/software simulation, to demonstrate how components will interact with each other and the end user. For most baseband equipment development projects, it is advisable to start with a known architecture that can be adapted to the specific needs. This has the advantage of providing a working baseband system, with components that perform functions and that can communicate with each other according to established processes and timing. On the other hand, any existing system brings with it certain built-in constraints that may or may not hinder the conversion to the new requirements. The difficulty here is that one cannot always make a complete determination on this issue until the development project nears completion.

6.3.2 Modulation Systems and Modem Design

The engineering principles behind modulation and modem design as related to satellite communications are extensive and have undergone rapid improvement in recent years. In his extensive work in this field, Dr. Fuquin Xiong of NASA has conducted an inquiry into this subject in which he observes that digital modems have become a cornerstone of the effective response of satellite communications to terrestrial networks [5]. The drivers are an efficient use of power and bandwidth, along with a demand for high service quality and reliability of transmission. We benefit from the thoroughness and clear presentation of his research.

The selection of a particular type and design of modem depends heavily on the particular modulation format, as summarized in Figure 6.6. Digital modulation can be subdivided into categories that have constant envelope (i.e., the envelope of the signal amplitude does not vary appreciably during transitions between bits or symbols) or nonconstant envelope (i.e., the envelope varies and in fact may collapse to zero during normal transmission). In general, constant envelope is preferred for nonlinear RF channels such as

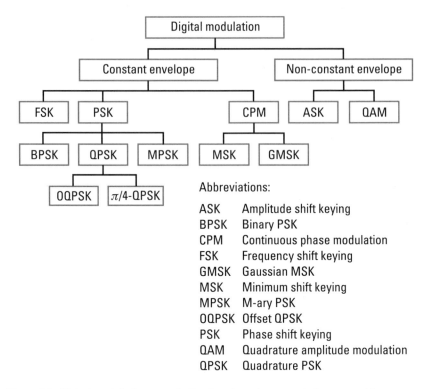

Figure 6.6 Digital modulation tree, indicating the primary techniques used in commercial satellite communications.

a saturated satellite transponder, whereas nonconstant envelope usually demands linear channels. Our examples to follow will be for forms of PSK that are popular in commercial satellite communications where nonlinear channels are commonplace. A generalized modem block diagram of a QPSK modem is provided in Figure 6.7, assuming constant bit rate transmission in a continuous mode. Significant modifications are needed to support a higher order modulation, e.g., eight-phase PSK, or to allow burst operation for TDMA. In Figure 6.8, we provide examples of unfiltered modem output spectra for binary PSK (BPSK), QPSK, offset PSK (OPSK), and minimum shift keying (MSK). The lower sideband levels of MSK, and to some extent OPSK, are a consequence of maintaining a more constant envelope than standard QPSK and BPSK provide. On the other hand, the main lobes of these particular modulation formats are broader. In general, these spectra would be further filtered at the modem output to reduce adjacent channel interference.

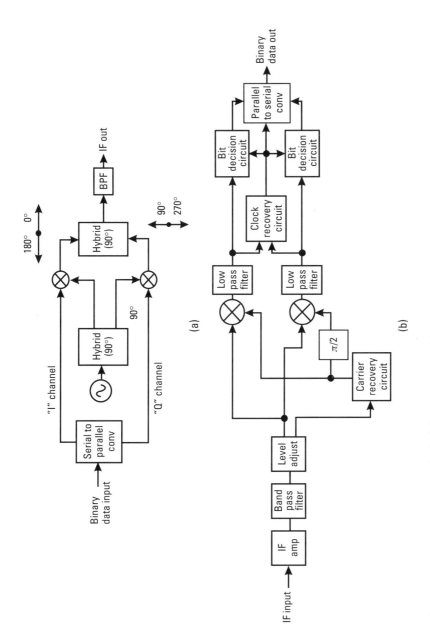

Figure 6.7 Generalized block diagram of a QPSK (a) modulator and (b) demodulator.

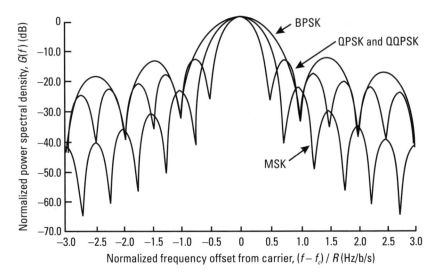

Figure 6.8 Normalized power spectral density for BPSK, QPSK, OPSK, and MSK.

The important determinants of modem link performance are as follows:

- Modulation format (according to Figure 6.6), based heavily on theory of statistical communications in presence of noise;

- Forward error correction (FEC), implemented with one of a number of standard error correction codes—convolutional encoding, turbo coding, block encoding, or a combination (i.e., concatenated coding);

- Decoding approach, such as soft decision or hard decision for convolutional coding;

- Techniques for carrier, bit, and timing recovery;

- Interference environment, including co-channel and adjacent channel;

- Use of spread spectrum modulation as in code division multiple access;

- Non-linearity effects in ground and satellite HPAS;

- Effect of the fading environment on modem operation.

It is impossible to provide the full range of possible performance under all of the variables. Examples of modem performance for a subset are provided in Figures 6.9 and 6.10 for convolutional and block error-correcting codes, based on theory. Each graph includes the performance of the uncoded digital channel as a reference. Concatenation (i.e., the sequencing of both block and convolutional coding) has proven extremely effective in the

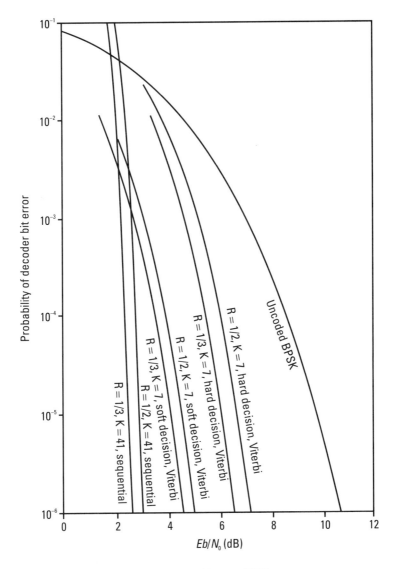

Figure 6.9 Convolutional code performance (*Source:* IEEE).

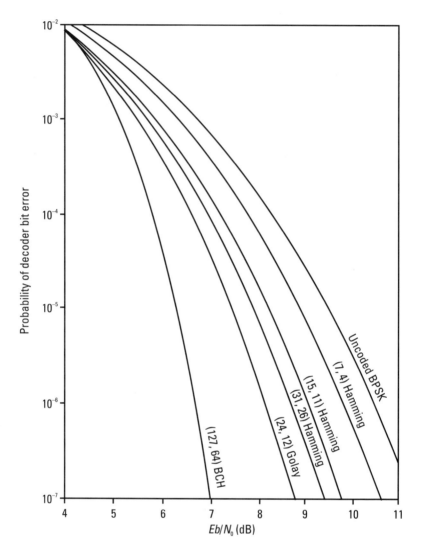

Figure 6.10 Block code performance (reprinted with permission of Prentice Hall, Englewood Cliffs, N.J.).

popular applications of DTH and GMPCS. Another trend is the application of more bandwidth-efficient digital modulation formats, such as eight-phase PSK (8PSK) and sixteen-level quadrature amplitude modulation (16QAM). These represent a tradeoff between bandwidth and power (i.e., additional power in terms of E_b/N_0 is needed to achieve the narrower bandwidth).

Characteristics of commercially available satellite modems are provided below. These are examples of stand-alone units that would be found in a major earth station used for medium- or high-speed data transmission. EF Data produces the SDM-300A satellite modem for data rates up to 2 Mbps. It utilizes proprietary digital signal processing, resulting in higher reliability and reduced packaging size. Table 6.2 provides a summary of specifications; its main features include:

Table 6.2
Summary of Technical Specifications of EF Data SDM-300A Satellite Modem

IF frequency range	50 to 180 MHz in 1 Hz steps
Digital interface	EIA-232, EIA-422, and V.35
Digital data rate	2.4 Kbps to 5 Mbps in 1 bps steps 64 Kbps to 5 Mbps 8PSK
Information rate	4.8 Kbps to 2.5 Mbps
Modulation/demodulation	BPSK (R = 1/2) QPSK (R = 1/2, 3/4, and 7/8) 8PSK (R = 2/3)
Plesiochronous buffer for timing adjustment	2 to 99 ms in 2 ms steps
Forward error correction	Viterbi, K = 7 (R = 1/2, 3/4, 7/8) Sequential (R = 1/2, 3/4, 7/8) Reed-Solomon per INTELSAT standard
Data scrambling	IESS-308 (V.35), IESS-309, or none
External reference input	1, 5, 10, 20 MHz
Agency approval	CE mark
Output power	−5 to +5 dBm, adjustable in 0.1 dB steps
Output spurious	5 dBc, 0 − 500 MHz (4 kHz band)
Output return loss	20 dB at 75
Data clock source	Internal (1×10^{-5}) or external
Input power	−30 to −55 dBm
Carrier acquisition range	35 kHz from 100 Hz to 35 kHz
Acquisition time	64 Kbps (R = 1/2), sec
Buffer clock	Internal, external, transmit, recovered RX
Plesiochronous buffer	16 to 256 kbits
Size, weight	$4.4 \times 48 \times 40$ cm, 4.9 kg
Operating temperature and humidity	0 to 50 C, up to 95% noncondensing

- 2.4 kbps to 5 Mbps;

- Flexible topology to simplify conversion for other features;

- Intermediate data rate (IDR) for public network services over INTELSAT;

- INTELSAT Business Service (IBS);

- Drop and insert (D&I);

- Automatic uplink power control (UPC);

- Asynchronous channel unit overhead;

- Reed-Solomon block encoding and decoding;

- Carrier and bit timing acquisition in less than one second;

- Built-in self test;

- QPSK and 8PSK modulation.

This modem is designed for a range of applications typically in medium to large earth stations. Units of this type have been deployed around the world as part of the INTELSAT system for public telephone services and in private networks (e.g., IBS) for medium- to high-speed data transmission. In all cases, the modem is operated on a continuous mode in an SCPC transponder. Additional modem transfer characteristics are provided in Table 6.3. In many applications, modem performance will vary widely and this information is representative of what real hardware is capable of doing under laboratory conditions.

This particular style of modem is intended for what might be termed industrial applications in telecommunications and data transmission. Devices of this sort are used for point-to-point links that allow Internet service providers (ISPs) outside of North America and Europe to gain access to the information services that those regions have to offer. Modem design has also been reduced to a very compact form for the user terminal, which will be discussed in Chapter 9.

Table 6.3
Bit Error Rate Performance of the EF Data SDM-300A Satellite Modem

Convolutional Encoder, Sequential Decoder				
Data rate	BER	E_b/N_0 (R = 1/2)	E_b/N_0 (R = 3/4)	E_b/N_0 (R = 7/8)
100 kbps	10^{-5}	4.8	5.8	6.7
—	10^{-7}	5.8	6.6	8.0
1.544 Mbps	10^{-5}	5.8	6.3	6.9
—	10^{-7}	6.6	7.1	8.0

E_b/N_0 for Convolutional Encoder, Viterbi Decoder				
BER	R = 1/2	R = 3/4	R = 7/8	8PSK, R = 2/4
10^{-5}	4.6	6.0	7.2	7.4
10^{-6}	5.3	6.8	7.9	8.2
10^{-7}	5.9	7.5	8.6	8.8
10^{-8}	6.4	8.0	9.4	9.6

Typical E_b/N_0 for Concatenated Reed-Solomon and Convolutional Coding				
BER	R = 1/2 (IBS)	R = 3/4 (IDR)	—	8PSK, R = 2/4
10^{-5}	3.2	4.0	—	5.8
10^{-6}	3.5	4.2	—	6.1
10^{-7}	3.6	4.4	—	6.4
10^{-8}	3.8	4.6	—	6.6

References

[1]   Viardot, Eric, *Successful Marketing Strategy for High-Tech Firms*, 2d ed., Norwood, MA: Artech House, 1998.

[2]   Urban, Glen L., and John R. Hauser, *Design and Marketing of New Products*, 2d ed., Englewood Cliffs, N.J.: Prentice Hall, 1993.

[3]   Fluckiger, François, *Understanding Networked Multimedia: Applications and Technology,* London: Prentice Hall, 1995.

[4]   Lee, Jhong Sam, and Leonard E. Miller, *CDMA Systems Engineering Handbook,* Norwood, MA: Artech House, 1999, p. 368.

[5]   Xiong, Fuqin, "Modem Techniques in Satellite Communications," *IEEE Communications Magazine,* August 1994, p. 84.

7

Earth Station RF Design and Equipment

The design of a typical earth station follows the principles of RF and microwave engineering, in the context of transmission of modulated carriers through the satellite to other earth stations. Because of the complexity of this task, we break the problem down into its constituent parts:

- *Uplink design*—microwave transmission toward the satellite, involving a high-power signal;
- *Downlink design*—the corresponding receive direction, involving low-power signals;
- *Antenna and microwave design*—the earth station components that interface directly with the RF path;
- *Upconversion and downconversion*—the portion of the earth station used to translate frequencies between IF and RF.

7.1 Uplink Design and the EIRP Budget

The transmit portion of an earth station is sized according to the EIRP required for the uplink path. The link budget, developed in Chapter 3, provides the starting point for RF chain analysis within the earth station; Chapter 8 gives an example of a PC software tool for convenient development on link budgets. Assuming that the uplink side of the link budget is defined, the designer needs to consider what can practically be implemented within the RF transmit portion of the earth station. There could be constraints on the maximum transmit power imposed by a limitation on prime power generation or radiation hazard (a particular concern in portable or handheld equipment). Also, there could be a size limit on the earth station antenna, such as for installation on the roof of a building or on a vessel or vehicle. Another consideration is the level of sidelobe radiation from the uplink antenna toward adjacent satellites.

The uplink portion of the link budget determines the required EIRP. The factors that constitute the required EIRP are the gain of the antenna, the loss of the transmission line (waveguide, coaxial cable [coax], or whatever is being used on the transmit side), and the output of the high-power amplifier (HPA) itself. The antenna could be a parabolic reflector of substantial diameter such as that used in a video uplink, or it could be a simple rod or helix antenna attached to a mobile terminal. We use the term *HPA* in reference to any type of amplifier; in actuality, the amplifier might only produce one watt of output (in which case it could hardly be called "high power").

The basic relationship for the EIRP is as follows:

$$\text{EIRP} = P_o - L_t + G_t$$

where P_o is the HPA output power at the waveguide flange; L_t is the transmission line loss between the HPA and antenna feed; and G_t is the antenna transmit gain at the operating frequency in the direction of the satellite. Starting with a required value of EIRP and uplink frequency, the designer determines the type and size of antenna, the type of transmission line (waveguide or coax, with its associated dB/meter loss factor), and the distance between antenna and HPA.

Working backward from the antenna,

$$G_t = \text{EIRP} + L_t - P_o$$

This simple relationship demonstrates that the gain we need increases with the required EIRP and the loss of the transmission line. Note also that increasing the transmit power has the direct benefit of allowing a reduction of the needed gain (evident by the minus sign).

7.2   Uplink Power Control

Depending on the requirements of the link, service, and space segment, the earth station may need some form of uplink power control (UPC) to reduce variation of power level at the satellite repeater. If we assume ideal operation without carrier level variation, the signals within a typical transponder may appear to be nonvarying over time. An example of this is shown in Figure 7.1(a), which represents an SCPC transponder with several low-power carriers and two higher-power carriers. Due to drift and/or link fading, the levels of the different carriers will change over time, producing the variability shown in Figure 7.1(b). This poses the following potential problems:

- The aggregate power of all carriers could drift upward, meaning that the total backoff of the transponder is reduced. This condition will generally result in greater IMD noise within the transponder and a reduction in the overall link C/N (assuming the link was previously operating at its optimum backoff).

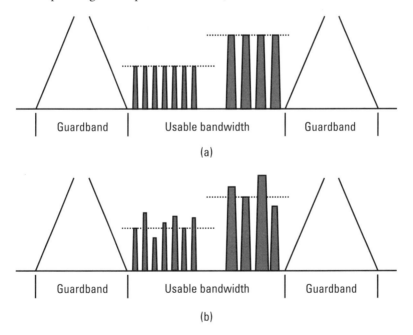

Figure 7.1 A multicarrier transponder: (a) with carriers aligned properly in power; and (b) with excessive power variations that degrade service for some and reduce the total backoff.

- A particular carrier that drifts downward in power will experience a direct reduction in uplink C/N of precisely the same amount, measured in dB. Also, the carrier will see a relatively higher level of IM noise, and the downlink C/N will decrease as well. Taken together, the link C/N will have been reduced through this combined mechanism.

- Uncertainty (error) in uplink level adds up to a mean value of service degradation. This must be compensated for by improving the overall link performance for all carriers. If carrier levels can be held within tight limits, then this margin is unnecessary.

- Uplinks may fade due to shadowing in L- and S-band mobile satellite systems or rain in Ku- and Ka-band fixed satellite services. As discussed in Chapter 3, additional power margin within earth station HPAs will be very useful for correcting for uplink fade.

There are two basic approaches for controlling uplink power variation: open-loop control within the earth station, and closed-loop control using a downlink beacon signal transmitted through the satellite. Open-loop systems sense the power within the uplink chain, usually at the output of the HPA and before the antenna. As shown in Figure 7.2, a small sample of signal power is taken from a directional coupler, detected and fed back to a controlled-variable attenuator at the input to the HPA. This compensates for internal drift and aging of earth station transmit components only. The closed-loop approach, shown in Figure 7.3, adjusts for equipment drift and uplink fading as the means to maintain nearly constant power at the satellite.

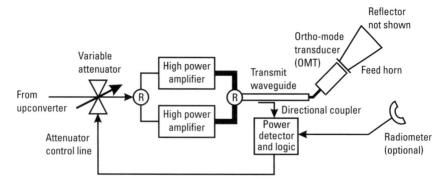

Figure 7.2 Open-loop uplink power control system, using an optional RF radiometer to detect rain attenuation from the ground.

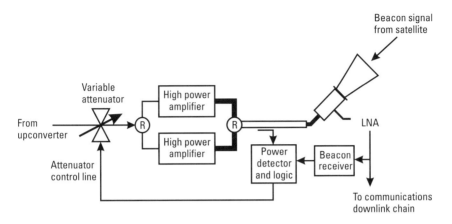

Figure 7.3 Closed-loop uplink power control system, using a dedicated satellite UPC beacon.

The satellite would be required to transmit a nearly constant downlink beacon carrier, which is either the normal telemetry carrier or a special signal provided within the payload for that purpose. Another, less accurate, approach is to use a carrier uplinked from one of the earth stations within the footprint. The obvious difficulty with an earth-station-generated beacon is that it, too, is subject to equipment drift and uplink fading. In either case, the earth stations with power control have a dedicated receiver to detect the average beacon level, and use this to control the variable attenuator ahead of the HPA. (In Chapter 6, it was suggested that the modem might perform this function as well.)

An important issue in UPC design relates to the response of the control loop to the dynamics of fading. This can be slow or fast, depending on the phenomenon, e.g., rapid fades due to multipath or slow fades due to widespread rain. It is not possible to anticipate every possible condition that could be experienced on the link; hence, the control loop may have to be adaptive in nature. Performance of the loop can be verified using link fading simulation, which can be performed by a special link simulator or an appropriately programmed computer.

UPC systems have been successfully applied to Ku-band BSS broadcast centers that originate digital multiplexed video channels, as well as hub stations for the outroute carrier used with VSAT star networks. Even though UPC is a dependable way to maintain network service availability during various forms of fading, it can go out of alignment due to normal wear and tear. Therefore, it becomes necessary to continuously monitor the operation of the equipment and perform preventive maintenance to assure that the controlled carrier levels are proper. Also, the satellite beacon carrier, which is the reference for closed-loop UPC, is a critical element and must itself be stable and dependable.

7.3 Transmit Gain Budget

The gain budget is very similar to the link budget except that it represents the internal RF equipment chain within the earth station itself. We use as an example the block diagram of a video-class earth station in the transmit direction, shown in Figure 7.4. Quite obviously, the particular type of equipment and its arrangement must be known (or at least understood in general terms) in order to put the block diagrams together and then create the gain budgets. We will evaluate the downlink gain budget, which is very similar, in the next section.

In the transmit direction, we see that the first stage is the exciter, a combined modulator-upconverter in one unit. A typical upconverter is capable of outputting its carrier at the desired RF frequency at a power level in the range of −60 to −20 dBm (e.g., −90 to −50 dBW). The particular value is determined by the required transmit power from the antenna and the intervening gains and losses after the upconverter. The corresponding link budget for the uplink at 18.2 GHz, given in Table 3.2, indicates a required EIRP of

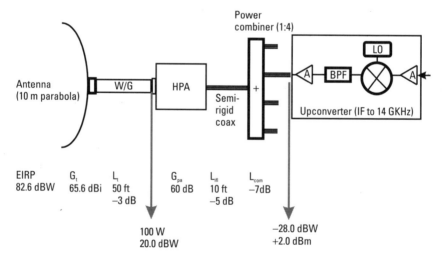

Figure 7.4 Uplink gain budget for an earth station high-power transmit chain (assumed to be at 18.2 GHz).

82.6 dBw. Assuming a 13-m antenna with 65.6 dBi of gain and 3 dB
of transmit waveguide loss, the HPA output power needed is 20 dBW (i.e.,
100 W). RF performance of the component is shown in Figure 7.4; i.e., the
gain of the HPA itself, the passive loss of the coax transmission lines internal
to the station, and the loss of the power combiner used to connect redundant
upconverters.

Values for these characteristics can be obtained from suppliers and
standard catalogs. The losses for the waveguide and semirigid coax cable are
taken from the typical data in Table 7.1 and Figure 7.5 [1].This indicates
that a waveguide at 18.2 GHz would have about 0.3 dB of loss per meter,
while the corresponding loss for semirigid coaxial cable is about 1.5 dB per

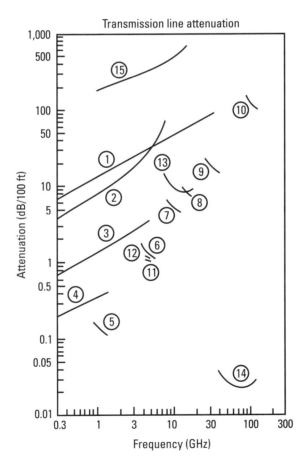

Figure 7.5 Attenuation characteristics of typical coaxial and waveguide transmission
lines.

Table 7.1
Properties of Typical Coaxial and Waveguide Transmission Lines (Figure 7.5)

Description	Material	Type #	Outer Dimensions (inches)	CW Power Handling	Flexibility
1. Coaxial	Teflon	0.141" dia	0.14 dia	50W	semiflexible
2. Coaxial	Polyethylene	RG 8	0.42 dia	30W	flexible
3. Coaxial	Helical polystyrene	7/8" HELIAX	1.0 dia	700W	semiflexible
4. Coaxial	Air	3-1/8" RIGID	3.5 dia	12kW	rigid
5. Rectangular WG	Aluminum	WR 770	8 × 4	57MW	rigid
6. Rectangular WG	Aluminum	WR 187	2 × 1	3MW	rigid
7. Rectangular WG	Brass	WR 90	1 × 0.5	730kW	rigid
8. Rectangular WG	Brass	WR 62	0.7 × 0.4	440kW	rigid
9. Rectangular WG	Silver	WR 28	0.36 × 0.22	95kW	rigid
10. Rectangular WG	Silver	WR 8	0.16 dia	1.8kW	rigid
11. Elliptical WG	Aluminum	RG 379	2.5 × 1.5	20kW	semirigid
12. Flexible Elliptical WG	Copper	WE 44	2.5 × 1.5	4kW	flexible
13. Ridged WG	Aluminum	WRD 750 D24	0.7 × 0.4	100kW	rigid
14. Overmoded Circular WG	Copper with Teflon liner	60mm dia	4 dia	—	rigid
15. Microstrip	Gold on Alumina	0.025"	—	50W	rigid

meter. Our 100-watt HPA has a nominal gain of 60 dB, which is typical for a complete unit with final amplifier and driver amplifier. The loss of a passive power combiner is basically the inverse of the number of inputs to be combined, which in this case is 1/4 (about 6 dB). An additional attenuation of 1 dB accounts for internal resistive loss, yielding a total of 7 dB for the four-way power combiner. Working backward, the output of the upconverter would have to be −33 dBm (−63 dBW) in order to produce the desired 20 dBW at the output of the HPA. It is customary to specify output of low-level devices such as upconverters and modulators in dBm. A discussion of various classes of HPAs and their power capabilities is provided in Section 7.6.

The uplink gain budget reveals whether all of the active and passive elements of the earth station are compatible with one another. For example, does the HPA have adequate input power to achieve the required input to the antenna? If not, then another driver amplifier needs to be introduced into the chain between the upconverter and HPA. Uplink power control, if provided, puts additional demands on the uplink for amplification and output power.

Assuming that there is enough drive to the HPA, the other critical aspect of meeting the EIRP requirement is the amount of loss between it and the antenna feed. This should be controlled as carefully as possible, as every 0.5 dB of loss represents a power reduction of approximately 10%. The types of items to be encountered include:

- *Transmission line* (typically waveguide or simirigid coax), as reviewed previously. The amount of loss is predictable and can be calculated based on the type of line and its length. Flanges or other connectors will introduce attenuation from power absorption as well as additional loss due to voltage standing wave ratio (VSWR), the latter governed by the formula in dB:

$$L_m = 10 \log \left(\frac{[VSWR + 1]^2}{4 \bullet VSWR} \right)$$

Environmental factors also play a role in changing the transmission characteristics of the line. One of the biggest concerns is contamination and the resulting rust and corrosion from air and moisture. This is why lines are fed dry air or nitrogen (the latter introducing a change in dielectric constant and hence a shift in center frequency of waveguide filters in the same line).

- *Power combiners*, typically using microwave filters to minimize insertion loss. This topic is reviewed later in this chapter. Other filtering might be included to reduce uplink out-of-band radiation (to protect other services such as space research and radio astronomy, discussed in Chapter 1).

- *High-power microwave switches* that allow backup HPAs to be connected in and for routing the uplink to alternate antennas. These switches are typically mechanical in nature so as to provide full bandwidth operation with minimum insertion loss. They should be installed in a building or shelter to protect them from the elements. For larger stations it is useful to consider RF switching and routing between HPAs or individual uplink channels, or even to allow these channels to be connected to any particular antenna on site. This arrangement would be controlled through the monitor and control (M&C) system of the earth station. An important point is that the particular routing of uplink chains to antennas should be made clear to operators at all times so that a mistake or failure does not cause harmful interference to another transponder or satellite.

7.4 Downlink G/T and RF Level Analysis

The function of downlink design is to meet the G/T requirement and deliver sufficient carrier power to the downconverter and demodulator. The G/T analysis considers the total received noise due to the antenna, waveguide, low noise amplification, and subsequent stages. This is evaluated against the net antenna gain as measured at the same reference point in the earth station RF receive section. The gain budget technique is then used to verify signal level in the same manner as for the uplink. These aspects are reviewed in the following sections.

7.4.1 Downlink Noise Temperature Allocation

In Chapter 3, we introduced the system noise temperature, T_{sys}, that defines the sensitivity of the receiving system for all earth stations. Noise is basically derived from the irreducible thermal noise due to the random motion of electrons in conductors [2]. Within the RF bands of interest, we have the basic relationship for the noise spectral density, N_0, in watts per Hertz (W/Hz):

$$N_0 = kT_{eq}$$

where T_{eq} is the equivalent noise temperature in Kelvin (K) and k is Boltzmann's constant. As measured across a resistor with resistance, R, and in a bandwidth, B, the resulting RMS voltage is given by Nyquist's formula,

$$V = \sqrt{4kT_{eq}BR}$$

We can assume that both the power source and load have the characteristic impedance, R, over the bandwidth of the carrier. Then, the noise power in the load for the bandwidth of interest, B, is simply

$$N = kT_{eq}B$$

For a typical arrangement of earth station RF receive components, the system noise temperature is

$$T_{sys} = \frac{T_a}{l_r} + \frac{l_r - 1}{l_r} \cdot 290 + T_{re}$$

where T_a is the antenna temperature; l_r is the loss factor ($l \geq 1$) for the input receive waveguide or transmission line; 290K is the assumed physical temperature of the waveguide; and T_{re} is the equivalent noise temperature of the receiver (including noise added bystages downstream of the LNA).

From this relationship, we see the criticality of controlling waveguide loss and using LNAs with appropriately good performance in terms of noise contribution. Antenna temperature for reflector antennas is generally under 50K for dishes larger than about 40 wavelengths (e.g., 3m at C band, 1m at Ku band, and 60 cm at Ka band). As shown in Figure 7.6, noise pickup at high elevation angles is nearly constant and is dominated by background noise from space and antenna backlobes. Lower elevation angles introduce more ground noise. Evaluations of this type have been attempted by performing an integration across the entire antenna radiation pattern in three dimensions. However, best accuracy is obtained by measuring the antenna temperature directly with a calibrated receiver.

The situation for very broad-beam antennas such as those used for mobile communications is that antenna temperature is often greater than T_{re}. As illustrated in Figure 7.7, a simple mobile antenna will allow ground noise to enter at nearly the same gain as the desired signal [3]. Summation of the

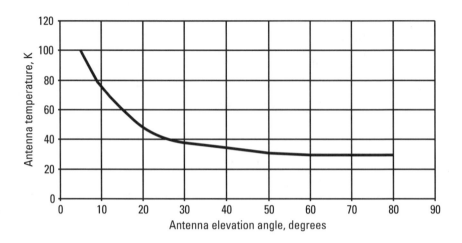

Figure 7.6 Typical antenna temperature contribution with elevation angle.

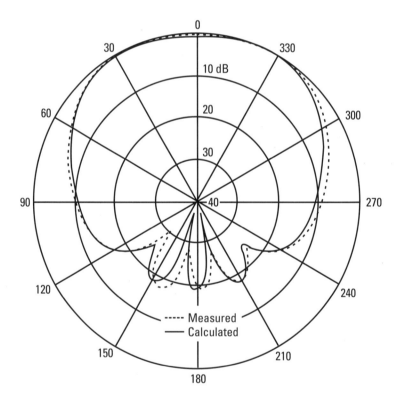

Figure 7.7 Broad beam pattern of a rectangular microstrip patch MSS antenna.

total noise throughout the broad antenna pattern is also a difficult analytical process because of coupling effects from adjacent objects which distort the pattern. More noise may be introduced by local sources such as a human being or an operating vehicle. Therefore, noise in mobile user terminals may be dominated by the pickup of their antennas.

The noise temperature of a particular LNA can be computed from the performance of its components. Figure 7.8 presents an example consisting of an input waveguide and flange, low loss isolator, two stages of low-noise GaAs amplification, a final high-gain stage using bipolar silicon transistors, and an output isolator. The function of the input and output isolators is to match the amplifier to the transmission lines. This produces the lowest equivalent noise and prevents reflected energy from distorting the frequency response. Amplifier noise temperature and noise figure are directly related to each other by the following:

$$T = (F - 1) \cdot 290$$

where F is the noise figure as a factor ($F \geq 1$) and 290 is the ambient temperature in Kelvin.

The basic relationship for determining T_{re} for a typical receiving chain is as follows:

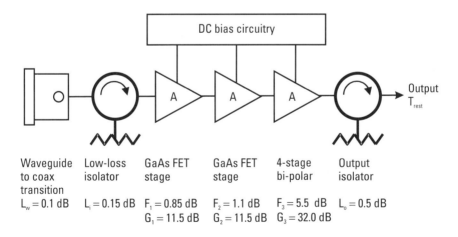

Figure 7.8 Low noise amplifier (LNA) block diagram, showing a waveguide-to-coax adaptor input, a three-port coax isolator, and three stages of amplification. The isolator at the output reduces interaction with downstream components.

$$T_{re} = (l_w l_i - 1) \cdot 290 + \left[T_1 + \frac{T_2}{G_1} + \frac{T_3}{G_1 G_2} + \frac{T_{rest}}{G_1 G_2 G_3} \right] l_w l_i$$

The first term represents the noise contribution of the input losses due to the coupler and isolator and the second term is the combined noise contribution of the stages of amplification. The factor, $l_w l_i$, increases the noise effect since it is due to the loss ahead of the active amplifier stages. The summary of the overall performance of this LNA is presented in Table 7.2. The total noise temperature of this LNA is approximately 100K, two-thirds of which is contributed by the first Gallium Arsinide field effect transistor (GaAsFET) stage. Note also the total gain, which amounts to approximately 54 dB. Both of these values are typical for Ku-band LNAs without additional noise reduction technology such as Peltier cooling.

We now evaluate the systems noise temperature for the earth station, consisting of the antenna, transmission line, LNA (or LNB, if appropriate), and the remaining stages of the RF/IF downlink system. An example of the analysis of system noise is provided in Table 7.3 for the receiving system shown in Figure 7.9. We are assumed to be looking into the LNA itself as the plane of reference. In calculating the G/T, it does not matter what the reference plane is as long as both G and T are evaluated at the same point.

Table 7.2
Analysis of T_{re} for Typical LNA String Shown in Figure 7.8

Component	Gain of This Stage, dB	Noise Figure, dB	Noise Temperature, K	Gain of Stages Preceding Stages, dB	Contribution, K
Waveguide/ isolator	−0.25	0.25	17	0.0	17.18
GaAs LNA stage 1	11.5	0.85	63	−0.25	66.41
GaAs LNA stage 2	11.5	1.10	84	11.25	6.27
Bipolar amplifier stage	32.0	5.50	739	22.75	3.92
Output isolator	−0.5	0.50	35	54.75	0.00
Total at output	54.25	—	—	—	93.78

Table 7.3
Analysis of T_{sys} for Receiving Earth Station in Figure 7.9, Referenced to Input to LNA

Element	Gain, dB	Noise Figure, dB	Noise Temperature, K	Gain With Respect to LNA Input	Contribution, K
Antenna feed	−0.25	—	40.00	0.25	37.76
Input waveguide	−0.25	0.25	17.18	0.25	16.22
LNA (input is reference)	54.00	1.22	94.06	0.00	94.06
Cable	−10.00	10.00	2610.00	54.00	0.01
Receiver	—	12.61	5000.00	44.00	0.20
System noise temperature	—	—	—	—	148.25

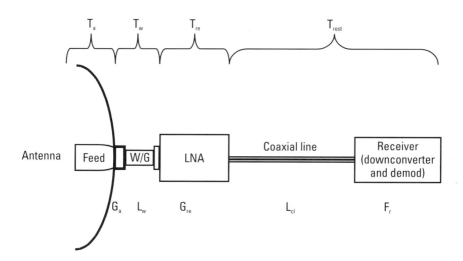

Figure 7.9 Typical block diagram to determine receiving system noise temperature.

7.4.2 G/T Budget

The evaluation of G/T at this point is very simple: it is the ratio of the gain to the system noise temperature (or difference in dB) at the same point in the receiving system (e.g., the input to the LNA). The antenna gain is reduced by the amount of waveguide loss already budgeted in Table 7.3 (e.g., 0.25 dB). Assuming the 45-cm-diameter antenna in the link budget from Chapter 3 in Table 3.1, the receive gain at the antenna feed output flange is 34.0 dBi. This is reduced by 0.25 dB, yielding 33.75 dBi of effective gain. Our corresponding value of noise temperature is 150K, or 21.8 dB ≠ K. The G/T of this receiving earth station is therefore 12.0 dB/K. This would correspond to a condition of (1) perfect alignment of the main beam with the satellite, and (2) clear sky at an elevation angle of greater than about 25°.

7.4.3 Gain Budget

The downlink gain budget is best understood through an example, which we take from the diagram in Figure 7.10 and the link budget defined in Table 3.1. The carrier level at the input to the LNA (e.g., the output of the antenna feed and waveguide) is calculated as:

$$P_r = EIRP - A_o + G_r - L_r$$

From Table 3.1, this is 52.0 − 205.5 + 34.0 − 0.25 = −119.75 dBW. The RF components in the receiving system introduce the gains and losses indicated in Figure 7.10, giving a received power of −83 dBW (−53dBm) at the input to the video receiver. We would like this power to be about in the middle of the receiver's design range. Assuming this range to be −40 to −60 dBm, we have a satisfactory design for the downlink. If this were not adequate for the downlink, then we have two choices: buy a receiver that meets our "middle of the range" criterion, or adjust the preceding gain

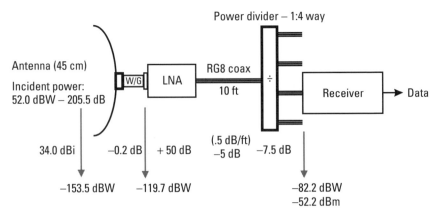

Figure 7.10 Typical receiving chain gain budget at 12 GHz.

accordingly. Increasing the gain usually means buying an LNA (or LNB, as appropriate) with more gain. If such a device is not available, then we need to insert a driver amplifier between the LNA and receiver. The noise contribution of this amplifier will be negligible for the level of LNA gain assumed in this example. Inserting a fixed value of attenuation can decrease gain. This might occur naturally if the coaxial transmission line is extended in length (or, alternately, a more lossy type of cable is used). Otherwise, we can insert a fixed or adjustable attenuator.

7.5   Antenna and Microwave Design

The objective of antenna and microwave design is to provide a nearly transparent path to the link for the baseband equipment and, ultimately, the end user. This is addressed through analysis of element performance based on its physical and electrical properties, governed by the laws of electromagnetic theory and microwave electronics. While the capability of these devices has improved as their cost has come down, the basic principles have not really changed very much over the 60 years of refinement of the technology base. A basic principle is that of guided waves through metal structures such as rectangular waveguide, coaxial cable, and various compact arrangements that come under the general category of stripline. These are covered in Section 7.5.1.

Another consideration relates to the consistency and reliability of operation over time, as the RF portion of the earth station is exposed to a potentially hostile outdoor environment. Rain, ice and snow, wind and salt spray, sun, and heat and humidity produce various forms of degradation, both short term and long term. For this reason, RF equipment must be properly protected, its performance monitored, and corrective action taken at the first sign of trouble. Repairs under these circumstances could take the facility out of service for a considerable time.

In the next paragraphs, we review the design principles for the essential elements, including antenna feeds, transmission lines, and RF filtering and multiplexing (principally on the uplink side). Antenna structures can be simple fixed designs such as those used for DTH services, complex tracking antennas or large-diameter uplink antennas in GEO and non-GEO systems. Readers needing more detail on the specifics of these designs can refer to the references at the end of the chapter. In any case, specific antenna designs will require the attention of the antenna manufacturers who are able to design and produce the mechanical and electrical portions and integrate them into a complete working assembly. This is a complex process and includes the care needed to properly align the system and perform antenna performance tests, including pattern measurements on an antenna range.

7.5.1 Waveguides and Transmission Lines

Microwave frequencies can propagate best within metallic tubular structures as opposed to pairs of wires, common in telephone and low-speed data applications. This is not to say that you cannot use good-quality twisted pair cable to transfer microwave signals; however, the substantial attenuation and poor external isolation of this type of cable renders it far less appropriate than the approaches to be discussed in this section. A selection of waveguide structures is provided in Figure 7.11, illustrating many forms of hollow waveguide, coaxial lines, and stripline [4]. The following properties are important to their application in an earth station:

- *Rectangular waveguide (hollow inside)*—the width, a, and the height, b, determine transmission characteristics, principally attenuation versus frequency. The principle property of rectangular guide is the lowest frequency that can propagate, e.g., the cutoff frequency of the dominant mode (discussed below).

- *Circular waveguide (hollow inside)*—the diameter, d, is the only dimension in question, likewise determining the cutoff frequency; a flexible elliptical cross-section waveguide similar in mechanical design to electrical "BX" cable is popular for low-loss, high-power transmit applications at C and Ku band.

- *Coaxial line (circular or square cross-section)*—this will propagate microwave and lower frequencies, effectively down to DC. If the frequency is high enough, then the lowest dominant mode can propagate as in hollow circular guide. The square cross-section version has been called squareax and may be formed with a center conductor held by dielectric spacers and sandwiched between two machined plates.

- *Stripline and microstrip*—bimetallic transmission line with a strip conductor suspended over a flat sheet of metal called a groundplane; alternatively, there can be two groundplanes with the strip in between. This technique is popular for compact structures, particularly within an equipment module or microwave integrated circuit (MIC).

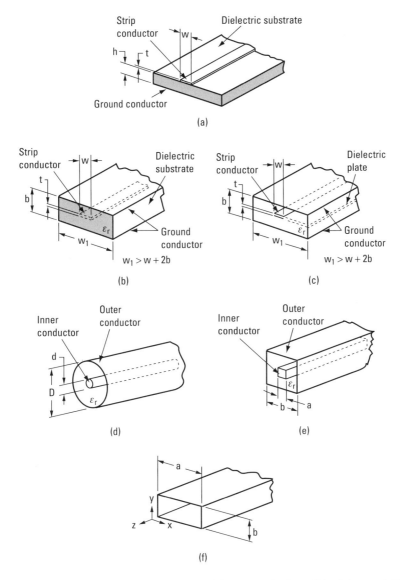

Figure 7.11 Microwave transmission lines: (a) microstrip line, (b) triplate stripline, (c) suspended stripline, (d) circular coaxial line, (e) square coaxial line (squarax), and (f) rectangular waveguide.

Propagation with low loss through the transmission line usually requires a very smooth interior surface and low electrical resistance (e.g., high conductivity, exemplified by aluminum, brass, copper, or silver, in increasing order of preference). The propagation of electromagnetic waves through typical transmission lines is illustrated in Figures 7.12 and 7.13 for rectangular and circular guides, respectively. Figure 7.12(a) displays transverse-electric (TE) modes in rectangular guide, wherein the electric component is oriented perpendicularly with respect to the direction of propagation. The top of the chart shows TE_{10}, the dominant mode with the electric component oriented vertically (indicated by vertical arrows). This propagates at the lowest possible frequency or, conversely, the longest possible wavelength, given by

$$\lambda = 2a$$

This simple formula conveys the fact that the wide dimension, a, must be larger than one half wavelength (this dimension must be adjusted for the actual wave velocity, which is less than in free space). As shown in (1) in the graphic, the arrows extend between the bottom and top walls, where the head is pointed toward the positively charged side. The horizontal dashed lines are the edge views of circular rings of the magnetic component. This is clearer in the top view of the guide, shown in (3). Along the length of the guide (2), the electric components reverse positive to negative (from pointing upward to pointing downward) to define the wavelength along the guide. One can think of a wave as having two components that are mirror images of each other, each taking a zigzag path parallel to the wide walls (i.e., bouncing off of the narrow walls of the waveguide).

We need to point out that the dimension, b, being smaller than a, cannot allow a horizontal wave to propagate properly. In effect, rectangular waveguide allows us to propagate TE_{10} in one polarization only. The only way to use a rectangular guide to carry horizontally polarized waves is to literally rotate it the required 90° so that dimension b is now vertical (not particularly elegant, but nevertheless the way it must be done).

To allow both horizontal and vertically polarized waves to propagate would require that a and b both satisfy the cutoff criterion of being at least one half wavelength. This is a role for square waveguide, often used in feedhorns (discussed in the following section). While this is useful in some applications, allowing propagation along both polarizations gives rise to the potential for moding (i.e., the transfer of energy from one into another, less desirable, mode). The second illustration in Figure 7.12(a) shows the first such higher order mode (TE_{11}) with curved electric and magnetic components. The cutoff wavelength is shorter than TE_{10}, demonstrating that it is, in fact, a higher order mode. Similar remarks can be made for TE_{21}, shown at

TE modes in rectangular waveguide

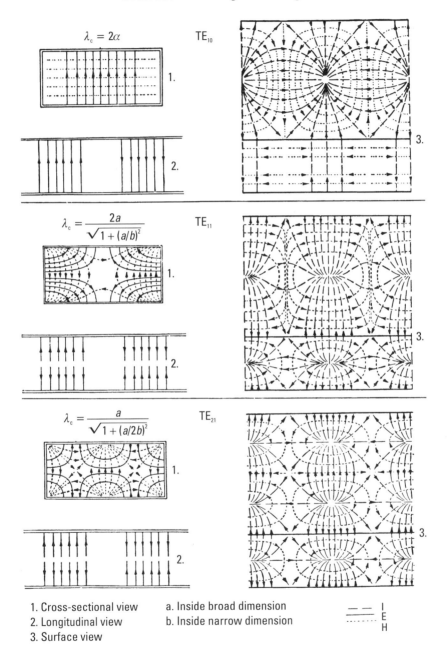

1. Cross-sectional view   a. Inside broad dimension
2. Longitudinal view   b. Inside narrow dimension
3. Surface view

Figure 7.12(a) Propagation modes in rectangular waveguide: TE modes.

TM modes in rectangular waveguide

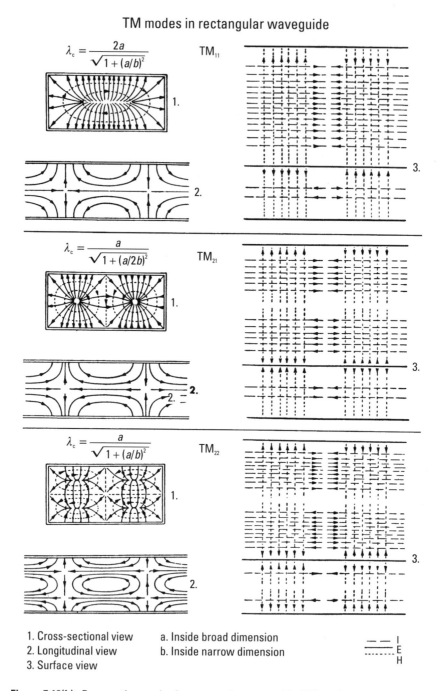

1. Cross-sectional view
2. Longitudinal view
3. Surface view

a. Inside broad dimension
b. Inside narrow dimension

Figure 7.12(b) Propagation modes in rectangular waveguide: TM modes.

TE modes in circular waveguide

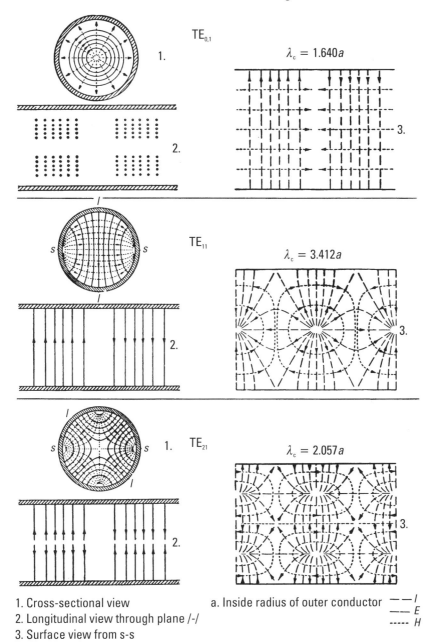

1. Cross-sectional view
2. Longitudinal view through plane /-/
3. Surface view from s-s

a. Inside radius of outer conductor

$-\,-\,I$
$-\,-\,E$
$-----\,H$

Figure 7.13(a) Propagation modes in circular waveguide: TE modes.

TM modes in circular waveguide

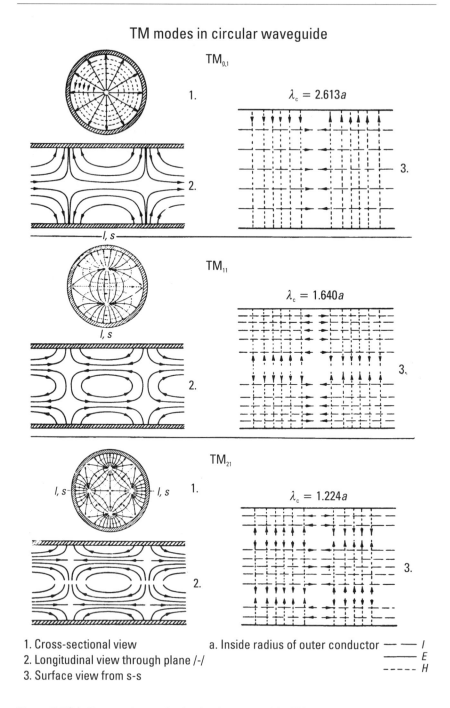

Figure 7.13(b) Propagation modes in circular waveguide: TM modes.

the bottom of Figure 7.12(a). Generally speaking, the only time when a microwave designer intentionally uses a higher order mode is within a specialized structure such as of a microwave filter or feedhorn.

Circular waveguide, shown in cross-section in Figure 7.13, has many applications in earth stations and demonstrates that, indeed, microwave mechanics is very close to plumbing. The TE modes in Figure 7.13(a) are related to their counterparts in rectangular guide. Here, TE_{11} is the dominant mode for use in earth stations for low-loss transmission associated with the antenna itself. The E vector in this example is vertically polarized; observe how the middle vector is straight, but those to the left and right are curved. In every case, the vector is perpendicular to the conducting wall where it touches, a boundary condition for electric fields in general. The fact that the opposite-sided vectors are symmetrical is good for propagation because it cancels out the nonhorizontal component. For this reason, an open circular guide can radiate pure horizontal polarization. The remaining examples in Figure 7.13 are provided for reference only.

7.5.2 Horns and Other Single-Element Antennas

Microwave propagation can be initiated through a radiating structure as simple as a piece of waveguide or wire of appropriate dimensions. These single-element antennas

- Provide a direct coupling of RF energy between the transmission line to space (the majority of antennas exhibit reciprocity, which means that the coupling is just as effective for energy from the transmission line to space as it is from space to the transmission line);

- Form electromagnetic energy into a beam to provide the desired directivity and possibly the suppression of sidelobes;

- Provide the correct polarization with the desired sense, e.g., linear (vertical or horizontal) or circular (left hand or right hand);

- Possibly allow the beam and polarization to be altered either mechanically or electrically.

These generic antenna requirements often represent tradeoffs with performance, physical size, cost, and convenience of use. Antennas are, after all, the most visible aspect of an earth station since they usually require an unobstructed view toward the satellite. Also, the physical ground below the antenna (which may be the earth itself, the top of a building or other structure, or the body of a vehicle, aircraft, or vessel) can affect the performance of the antenna. Generally speaking, the narrower the beam, the less interference we expect from the ground due to reflected RF energy and blockage.

We start with a discussion of elemental antennas, which are the simplest in structure and are often used in combination to produce more effective radiation characteristics. The most basic types of antennas are simple extensions of the end of the transmission line, such as the open end of the waveguide itself. The wave will be linearly polarized in free space because this is the same polarization within the guide, a principle often used in antenna design. The common problem with a simple waveguide antenna is that there is an impedance mismatch with free space, resulting in significant reflected power and a loss of radiation efficiency. The radiation pattern of the open-ended waveguide may be unacceptable for the application as well.

The way to both correct the impedance mismatch and create a more appropriate beam pattern is to flare the end of the waveguide in the form of a pyramidal horn, illustrated in Figure 7.14 [9]. Horns, in general, are desirable as feeds for parabolic reflector antennas, as will be discussed in Section 7.5.3. As shown by the vertical arrow, TE_{10} mode produces linear polarization while the flare angle and aperture opening determine the gain and impedance match to space. To change the sense of polarization, one only has to rotate the horn 90°, a process that requires a corresponding twist in the input waveguide. We could have used a square waveguide and horn to allow

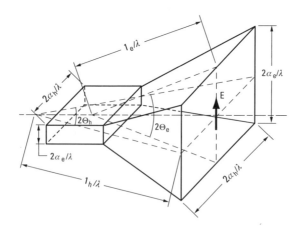

Figure 7.14 Configuration and dimensions for a smooth-walled pyramidal horn. The radiation pattern can be adjusted with the values of θ_e and θ_n. (Reference [9], with permission.)

both senses of polarization to operate at the same time. In this case, the waveguide would be capable of TE_{11} mode, which raises the possibility of undesirable moding. Another type of dual-polarized horn is the circular conical horn, which uses round instead of square guide. This also involves TE_{11} mode and offers somewhat improved performance in terms of circular polarization (CP) purity, sidelobe control, and manufacturing ease.

An important enhancement to improve the polarization purity of the conical horn is to introduce corrugations, shown in Figure 7.15. The corrugation helps purify polarization as well as control sidelobe levels. As with the preceding horn designs, the angle of the flare is used to adjust the impedance match and beam pattern. The refinement here is the introduction of circular ridges in the mouth of the horn. The spacing and height of the ridges produce higher order modes that help cancel some of the undesired radiation from the horn. An extension of this well-established design is the flanged waveguide horn (Figure 7.16), which contains a more limited number of circular (annular) slots. Horns of this type are commonly used in low-cost C-band reflector antennas.

Other classes of transmission line antennas are configured from cable with the ends exposed to space in an appropriate way. Familiar at HF (e.g., short wave bands between 3 and 30 MHz), they are made by pulling apart the individual conductors of twin lead or coaxial cable to form a half-wavelength dipole. Directivity can be improved by placing another wire behind the antenna to create the simplest form of Yagi-Uda array. Additional approximately half-wave elements placed in front of the dipole (called directors) further enhance the gain). At microwave frequencies, a dipole element formed by a slot in a metal plate or waveguide (illustrated in Figure 7.17). Multiple elements in an array may be driven in phase with the same signal to produce a focused or shaped beam, which is common in spacecraft antenna systems. Figure 7.18 provides waveguide implementations of slot arrays that produce focused fixed beams. With elements tied to individual amplifiers and phase shifters, it is possible to shift the beam in space by varying the electrical parameters (e.g., an electrically steered array) without physically moving the antenna structure. Such designs are being developed for application with non-GEO broadband systems and for mobile terminals.

Circular polarization (CP) is somewhat more complex to create but subsequently eliminates the need to perform mechanical rotations (since the wave is rotating). There are two approaches for creating CP in the first place.

- Introduce a discontinuity within a square or circular waveguide to split the LP wave into two components, and then delay one

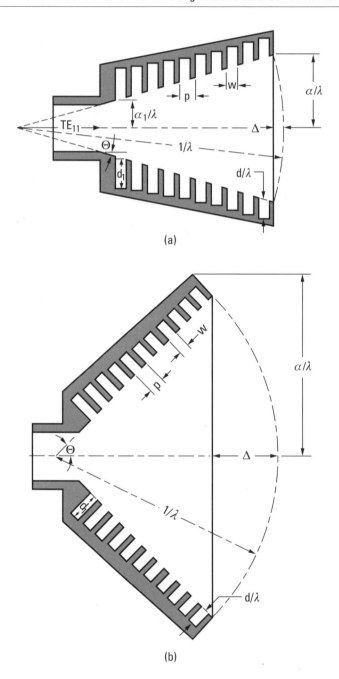

(a)

(b)

Figure 7.15 Cross-section view of corrugated horn showing usual slot configuration for (a) θ small, and (b) θ large. This design offers a symmetrical pattern with low sidelobes and high efficiency. (Reference [9], with permission.)

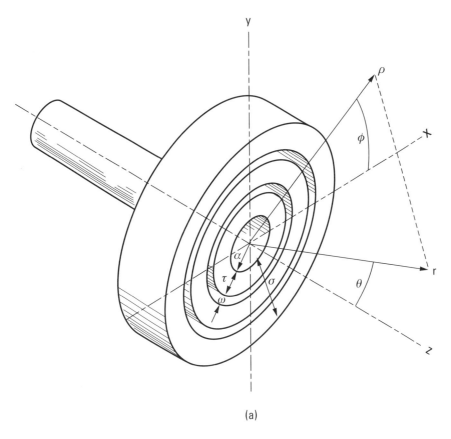

(a)

Figure 7.16 Compact circular flanged horn with annular slots cut into the waveguide. The slots are used to shape the pattern and reduce sidelobes. (Reference [9], with permission.)

with respect to the other by the equivalent of 90° of phase (shown in Figure 7.19) [5].

- Split the power equally and couple one of the inputs at an angle of 90° with respect to the other in an ortho-mode transducer. However, the orthogonally coupled signal must also be delayed by 90° so that CP is created (as in the previous case).

The advantage of the first approach is its simplicity, since it is performed within the waveguide and does not require added mechanical hardware. The second approach is completely mechanical and may offer greater

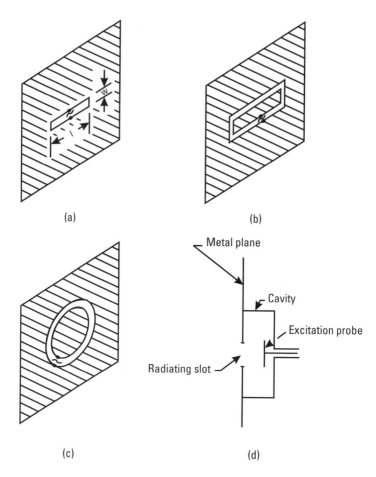

Figure 7.17 Slot antenna elements: (a) rectangular, (b) folded rectangular, (c) annular slot, and (d) cavity backed slot. (Reference [9], with permission.)

bandwidth and provide a more stable system. It is, however, more expensive to construct and is much bulkier in appearance.

7.5.3 Reflector Antennas

The basic properties of and requirements for reflector antennas such as the parabolic reflector are covered in our other work [6] and further investigated in another reference [7]. We concern ourselves here with the reflector antennas commonly used in earth stations for satellite communication. These are the familiar dishes used for video uplinks, mobile telephony gateways, VSAT

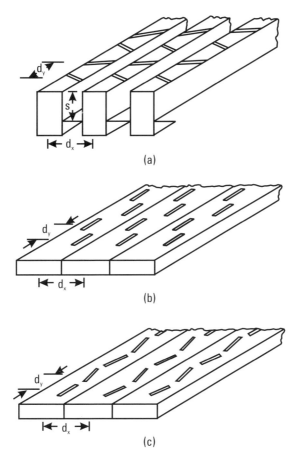

Figure 7.18 Waveguide slot arrays: (a) edge slot array; (b) displaced longitudinal slot array; (c) inclined series slot array. (Reference [9], with permission.)

hubs, and DTH receivers. The design of the reflector and feed follow principles that have not changed in decades, but are subject to numerous improvements that make them easier to install and operate, and reduce manufacturing costs. As readers are familiar, in reflector antennas size matters, since gain (as a true ratio) is proportional to area. Likewise, the beamwidth is inversely proportional to diameter.

Earth station antennas fall into a common set of alternatives as far as how the feedhorns are placed relative to the reflecting surface. In low-cost center-fed designs, the feed would be supported in front at the focus of the reflector; alternatively, for better performance and ease of maintenance,

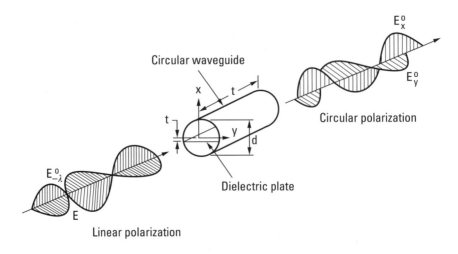

Figure 7.19 Conversion from linear to circular polarization by a 90° polarizer consisting of a dielectric plate.

the feed is placed behind the reflector and a subreflector is used to properly direct the internal beam. This approach is either called a Cassegrain or Gregorian, based upon the relative distance and shape (convex or concave, respectively), as shown in Figure 7.20.

There are a number of intriguing and potentially valuable reflector concepts that have been introduced for large earth stations. First demonstrated in the 1970s by COMSAT Laboratories, the torus antenna is a multibeam design that gets its name from the shape taken by its oblong reflector. As shown in Figure 7.21, the reflector is a segment of the surface of a toroid formed by rotating a single parabola about a vertical axis that passes through its focus. The reflector looks like a movie screen, and in fact screens of this type are used to increase the light intensity from TV projectors. As shown in the diagram, this geometry allows several feeds to be placed side by side without physically or electrically interfering with one another. Each feed sees approximately the same surface curvature; thus, there is no additional degradation normally associated with skewing a feed from the center line. The beams are produced in precise alignment, to be directed at different satellites along the geostationary arc. With the reflector oriented along the orbital arc, each feed can be placed for appropriate beam-to-satellite alignment.

Large antennas that must track MEO or LEO satellites can be quite cumbersome as far as how the energy is transferred between the feedhorn and the transmission line to the RF electronics.

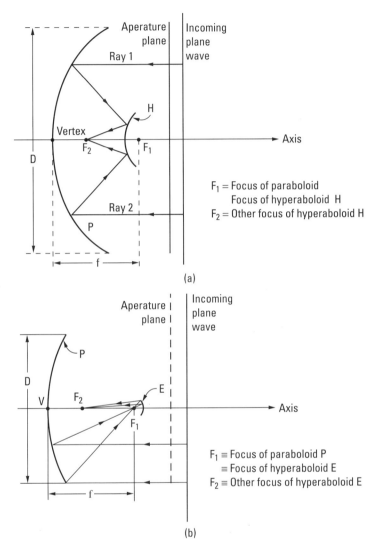

Figure 7.20 Geometry of the reflector and subreflector in an earth station antenna: (a) Cassegrain geometry, (b) Gregorian geometry.

We turn now to a discussion of the physical and operational properties of reflector antennas, since these aspects often determine how the earth station and overall ground segment will satisfy system requirements. Table 7.4 summarizes the more critical of these concerns, broken down into the major areas of electrical performance, mechanical properties, and system

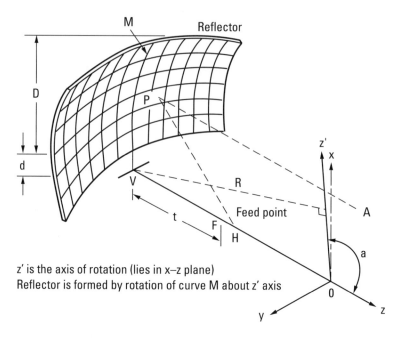

z′ is the axis of rotation (lies in x–z plane)
Reflector is formed by rotation of curve M about z′ axis

Figure 7.21 Multibeam torus antenna geometry.

requirements. Each key category is reviewed in a subsequent section below. While this list is lengthy, it cannot consider every important requirement and approach for a particular application. It is a checklist to show the extent of the issues involved, since the antenna is the one physical item that determines how well the RF link will perform—at the time of installation and throughout the seasons and operating life of the system.

7.5.3.1 Electrical Performance

The parameters of interest (gain, directivity, polarization, sidelobes, etc.) represent the input/output characteristics that determine how well the antenna will satisfy the requirements stated in a link or gain budget. Most of these can be measured at the factory or on an appropriate antenna range using standard RF test equipment. The actual installations can have considerable impact on these characteristics, which is why we consider the mechanical and system aspects in the next two subsections.

7.5.3.2 Mechanical Properties

Antennas are electrically conducting mechanical structures that radiate when excited by electrical energy. It is this duality that forces earth station design

Table 7.4
General Requirements for Earth Station Reflector Antennas

Characteristic	Considerations	Design Approach
Electrical performance		
Frequency	Link requirement	Straightforward feed design
Bandwidth	Frequency dependent	By design
Gain	Link requirement	Straightforward design
Antenna temperature	G/T	Inverse with size
Directivity pattern	Overall system performance	Determined by many factors
Cross polarization (axial ratio)	Interference design	Straightforward design, but dependent on installation quality
Power handling capability	High power uplinks	Waveguide and feed design
Port-to-port isolation	Simultaneous transmit and receive, and dual polarization	Waveguide and feed design
Out-of-band emissions	Compatibility with other radio services, e.g., radio astronomy	Proper filtering and feed design
Mechanical properties		
Angular travel	Ability to cover the operating segment of the GEO arc or sky	Requires appropriate mechanical mount
Ability to repoint without disassembly	Possible need to change alignment for other slots or orbits	Requires appropriate mechanical mount
Drive speed and acceleration	Critical for tracking non-GEO satellites	Motor power to keep antenna pointed at satellite position
Pointing and tracking accuracies	A function of antenna beamwidth and link budget requirements	Requires motor drives and signal tracking system; possible need for radome
Wind loading (operating and storage)	Heavy wind locations, including hurricanes and typhoons	To allow operation under certain wind, and safe stowage; possible need for radome
Other environmental requirements	High humidity and/or rain; low temperature and ice, corrosion and sunlight	Provide protection and blowers; provide heaters; possible need for radome
Reflector surface accuracy	Achieve the desired gain under all conditions	Appropriate reflector structural design and alignment; possible need for radome

Table 7.4 (continued)

Characteristic	Considerations	Design Approach
Mechanical properties (continued)		
Weight	A factor for mobile and transportable installations	Use lightweight materials like plastic, graphite, and titanium
Mounting requirements	Depends on location, fixed or mobile/transportable	Requires study of situation; may include detailed structural analysis
System requirements		
Functional requirements	Satisfy the local and systemwide requirements	Requires viewing the antenna as part of the overall system
Remote operation	Limited human intervention	Possible requirement for remote control operation
Maintainability	Will depend on the location and system requirements	Must be considered before specification and purchase
Lifetime	Depends on the application	Large antennas typically designed for 10 years or more; low-cost home installations may be up to 5 years
Interfaces (electrical and mechanical)	Transmission line connections, other power or utility connections to meet environmental needs	Must be considered before specification and purchase

engineers to understand many aspects of mechanical and civil engineering. Requirements for angular travel and motion of the reflector and feed derive from system requirements to keep the antenna pointed at the desired satellite. The mount and support structure would have the necessary degrees of freedom to accomplish this, typically through one of three mechanical arrangements (see Figure 7.22):

1. Azimuth-elevation—the most common approach (taken from the gun mount in artillery) can point the antenna anywhere in the sky, regardless of the orbital geometry. These have been improved over the years to give full coverage, using a central bearing structure that allows the system to rotate in azimuth the full 360° of the compass.

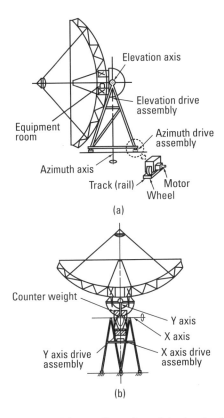

Figure 7.22 Antenna mount and drive configurations: (a) azimuth-elevation, (b) X-Y.

A possible disadvantage of the "az-el" mount is that it has a keyhole effect when following a spacecraft that passes directly overhead. At that time, the antenna reaches its limit at 90° elevation (e.g., zenith pointing) and must be rotated 180° in azimuth so that elevation can decrease to permit the beam to follow the satellite downward again from zenith.

2. Polar (not shown)—an arrangement used in most telescopes that directly compensates for much of the rotation of the earth. The primary axis of rotation is aligned with that of the earth, i.e., pointing toward Polaris, the North Star. These were popular for early use with GEO since most antenna adjustment is made in one axis to direct the beam at a different satellite.

3. X-Y—a system optimized for small changes about a fixed direction that also does not have a keyhole effect. Not attractive for most

situations where the antenna would have to be repointed frequently across the GEO arc.

A tracking antenna automatically keeps its beam pointed at the satellite using electric motors that drive the mount from position to position. The motors and servo loops must have the power to overcome inertia and friction throughout the travel, which can be a challenge for large structures and, of course, fast-moving satellites. Drive speed refers to the ability to keep the antenna moving at a constant angular rate, where acceleration is the more difficult challenge of overcoming the inertia of a possibly massive reflector and feed system. If RF electronics are also attached to the driven structure, then this becomes part of the inertial load of the drive motors. It is possible to include a set of redundant motors to provide backup in case of failure, something that should be considered if the operation of one particular antenna is critical.

The other aspect of tracking is the technical approach for controlling the drive to keep the antenna beam on the satellite as it moves along its orbit track. There are three basic methods for accomplishing this:

1. Program track—an open-loop control system that uses prerecorded ("canned") tracking instructions to command the antenna to move along the predicted path of the satellite. See Figure 7.23 for an illustration of a program track system for a shipboard antenna.

2. Step track—a simple closed-loop antenna pointing system that drives the antenna to a point of maximum received signal. As shown in Figure 7.24, the controller uses the amplitude of a received carrier (either a stable communication signal or a beacon

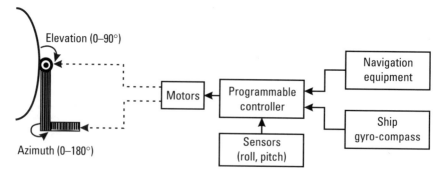

Figure 7.23 Functional block diagram of a program tracking system for a limited motion reflector antenna installed on a ship.

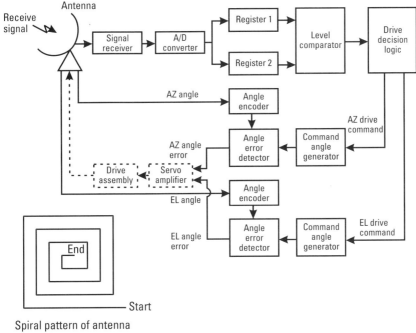

Figure 7.24 Block diagram of a step track system, which detects changes in received power and moves the beam according to an inward spiral pattern.

carrier generated onboard the satellite) as the only reference. The antenna is stepped in small increments typically producing 0.5 dB of change to hunt for near-optimum pointing using successive approximations. An example of the type of path is the rectangular inward spiral shown in the figure.

3. Autotrack (also called monopulse track)—a sophisticated closed-loop control system usually using both axes to sense error in pointing. As shown in Figure 7.25, these error signals are processed in the controller to be used to control the pointing of the antenna and beam. The action of an autotrack system is nearly continuous in maintaining constant alignment within the bounds of acceptable angular pointing error. The term *monopulse* refers to the use of a four-element feed system that provides sum and difference reference signals, by measuring a coherent downlink signal.

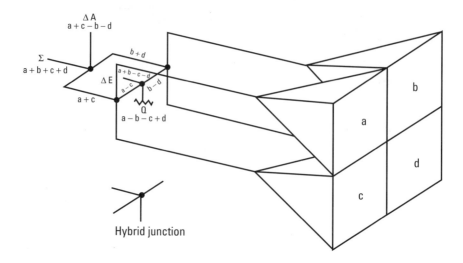

Figure 7.25 Simple two-axis monopulse feed.

7.5.3.3   System Requirements

The earth station antenna, particularly those of large size and with the ability to track satellites, must be specified in terms of the end use or application. These are system requirements, either from the standpoint of the operation at particular locations or, on a larger scale, the function within the context of the entire system or service environment. We have already addressed the electrical and mechanical performance parameters, which relate to the real-time aspects of antenna usage. Other requirements that can make a big difference for the lifetime of service include the following:

- *Functional requirements.* Included here are aspects of antenna operation that impact the overall system. These extend beyond antenna gain, pointing and tracking, and ability to redirect toward a different section of the orbit or sky. Examples include transportability, which is the ability to remove the antenna quickly, transport it by a specified means of transportation such as aircraft or vehicle, and reassemble the antenna for use at another location. Devices that can be transported and erected on site, like the one shown in Figure 7.26, are designed and manufactured with this need in mind. Another example would be an antenna installed in a commercial aircraft to provide bidirectional communications with a GEO satellite. Obviously, the antenna should be as flat as possible so as not to impact

Figure 7.26 Transportable/mobile video uplink antenna system (photo courtesy of Advent Communications).

the aerodynamic properties of the aircraft. A preference would be for a flat array that can track the satellite with an active beam of some sort.

- *Remote operation.* An antenna may be required to perform without human intervention or, alternatively, through a remote control network of some type. The latter could be over the satellite, which is acceptable as long as the antenna is already adequately aligned to provide the link. Redirection of large antennas is used in program distribution to TV stations, which is possible because only a few satellites might be used during a particular day. Operations personnel at a central monitoring center could activate the RF transmission, permitting remote adjustment of power level and polarization. The ability to shut down an errant uplink antenna system would be a necessity, since there would not be someone on site to inhibit the improper operation of the RF equipment or antenna system.

- *Maintainability.* In practice, this is extremely critical to the system operation and its ability to serve users in a reliable and consistent manner. The best example involves a gateway earth station for a GMPCS system. Extra antennas are included to allow for planned and emergency maintenance without reduction in availability. Another aspect that cannot be overlooked is the possibility of difficulties with the electronics for tracking the satellites. These systems can malfunction, driving the antenna off of a satellite, or precluding satellite acquisition in the first place. Lastly, there are human factors to be considered, in that it should be possible for maintenance personnel to perform these functions without risk to life or to the operation. Coming into play here are any local environmental or physical conditions that people could expect to experience, including wind, rain, and difficulty of reaching critical parts of the antenna system.

- *Lifetime.* Antennas can be designed and built to work for an extended period, often 10 years. Some can be counted on to last beyond the expected lifetime of a particular satellite or service. On the other hand, antenna designs using lightweight materials such as plastic cannot withstand harsh environments for more than a few years. The same can be said of electronics that are immediately attached to the antenna, including passive devices such as feeds, polarizers, and reject filters. This can tie back into the maintenance as discussed in the last paragraph, since almost anything can be made to last a long time if it is properly looked after (the converse is true as well, particularly when it comes to antennas). The local environment has a lot to do with this. Tropical climates take their toll through moisture (which produces rust and invades microwave plumbing), fungus (which literally eats its way through paint, rubber, and other organic materials), and high winds. Similarly, arctic latitudes are very difficult for antenna systems because of wide temperature swings that cause cracking of materials, and ice and snow can reduce performance and distort reflectors and support structures, which is why antenna deicing using electric or gas heaters is the standard in all but tropical and semitropical climates.

7.5.4   Main Beam, Sidelobe, and Cross-Polarization Performance

Some of the general performance characteristics applicable to a variety of antennas are summarized in Table 7.5. We have chosen antenna configurations that have been applied in satellite and terrestrial radio communications, particularly reflector and horn antennas. Planar array antennas were offered

Table 7.5

Comparison of Performance Characteristics for Earth Station Antennas

Property	Reflector	Horn	Planar Array	Hemispheric	Dipole and Yagi-Uda
Main Beam					
Peak gain	Determined by reflector size and efficiency; reduced by surface accuracy	Determined by aperture opening and horn design	Determined by type and number of elements, and manner in which array is combined	Determined by element design, groundplane configuration, proximate objects	2 dBi for dipole; between 6 and 20 dBi for Yagi, depending on number of elements
Beamwidth	Half-power points on main beam; determined by diameter and feed illumination	Half-power points on main beam; determined by horn geometry and design	Specified based on application; determined by element design and combination of elements	Specified based on application; determined by element design and surrounding structure	Toriodal for dipole; Yagi beamwidth determined by number of elements
Bandwidth	Determined by feed design	Determined by horn design	Determined by element design, array geometry, and power/phase distribution	Determined by element design	Applied at VHF and UHF
Sidelobe					
Sidelobes close to main beam	Determined by geometry and horn illumination	Determined by horn design	Determined by array geometry and power/phase distribution	Generally not present, due to size of main beam	For Yagi, sidelobes determined by number of elements
Sidelobes at considerable angles	Determined by geometry and blockage (feed and struts)	Typically substantially below main beam	Created by stray reflections and other aberrations	Created by stray reflections and scattering	For Yagi, pronounced back lobe
Cross-polarization					
Main beam	Determined by feed, OMT design, and reflector geometry	Determined by horn design	Determined by element design, array geometry, and power/phase distribution	Determined by element design and local reflections	Single element dipole or Yagi has good LP performance; CP established with crossed elements
Sidelobe region	Determined by horn design, geometry, and blockage (feed and struts)	Determined by horn design	Determined by element design, array geometry, and power/phase distribution	Generally not present due to size of main beam	Determined by number of elements

in the United Kingdom and Japan during the early rollout of DBS services, but more recently have been proposed for Ka-band applications. These could offer transmit/receive service and the ability to have a beam that follows LEO or MEO satellites. The hemispheric design is intended for GMPCS service, since it can view the entire sky even if the orientation of the antenna is not perfectly vertical. In the last column, we have the dipole and Yagi-Uda array antennas that are so common at VHF and UHF frequencies (e.g., for TV and FM radio broadcasting). Antennas of this type are applied for LEO satellite systems like Orbcomm, which operate below 1 GHz and provide low- to medium-speed data services.

A problem in specifying antennas reliably is the difficulty of making proper RF measures of these parameters. Physical size may make it very difficult to orient a reflector so that an appropriate signal at the desired frequency can be either received or transmitted in the far field region. A common practice is to use the same mechanical design as an antenna that has been fully characterized, then erect the new antenna on site. Once in working order, patterns can be measured (either fully or partially) to establish that (1) the antenna is assembled correctly (something that often is not the case on first check) and (2) not seriously affected by the local environment and terrain. A typical set of these measurements for a 10-m reflector antenna at C band is presented in Figure 7.27.

The pattern characteristics of horn antennas are perhaps the most predictable, based on theory and range measurement. This is because of the compactness and consistency of horn geometry (unless we are speaking of a very large horn that is composed of many pieces that may not have sufficient dimensional accuracy or stiffness). Horns have become more sophisticated over the years as designers have sought to improve efficiency and bandwidth; these improvements also benefit reflector antenna systems where horns are used as feed elements. The element of an array poses other issues, since by itself it behaves as a broadbeam antenna like that used for hemispherical coverage.

Antenna polarization properties provide isolation for frequency reuse in FSS and BSS satellite links. Large earth station antennas must provide between 35 and 40 dB of cross-polarization isolation within the main beam. The higher value is usually associated with linearly polarized feeds because of the greater simplicity of the hardware and measurement procedure. Polarization isolation is primarily the responsibility of the feedhorn, with the reflector and support affecting the performance only slightly. Achieving high isolation values in practice is a matter of good design and manufacture of the feed and polarizer; however, very-low-cost antennas may not be satisfactory

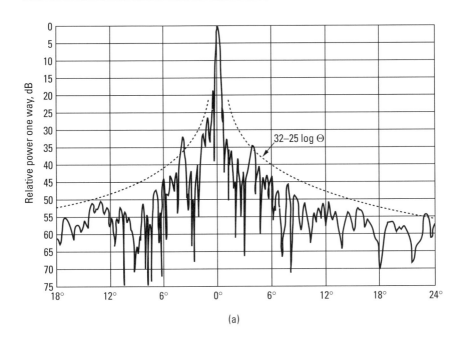

(a)

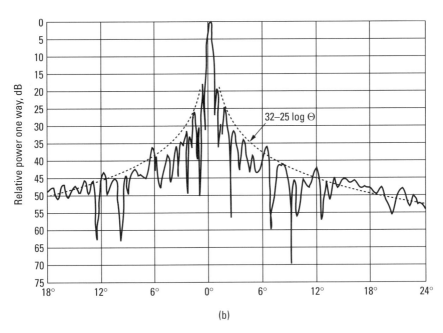

(b)

Figure 7.27 Measured 10-m earth station antenna pattern at C band: (a) transmit, 6,175 MHz, (b) receive, 3,950 MHz.

for demanding situations such as hostile physical environments, millimeter-wave frequencies, and networks with high internal or external RF interference. For example, the cross-polarization interference potential of the antenna affects transmissions on the cross-polarized channel or on an adjacent satellite that is cross-polarized in order to improve frequency reuse.

The surface of the reflector will introduce some distortion to the antenna pattern and potentially a reduction of gain and increase in sidelobes as well [8]. Figure 7.28 presents the loss of gain due to finite surface error. Another consideration is the presence of other objects or surfaces that the electromagnetic wave must pass through or around. For example, a radome may be employed to protect the antenna and RF electronics that must operate in extremely low temperatures or gusting wind. While it provides a comfortable working environment, it nevertheless introduces a direct loss of possibly less than 1 dB, as well as the potential for altering sidelobe levels. Other causes of concern are local structures and terrain features that introduce multipath reflections.

7.5.5 RF Filtering and Multiplexing

Large earth stations, such as those used as broadcast centers and international gateways, are required to transmit carriers in different transponders at the

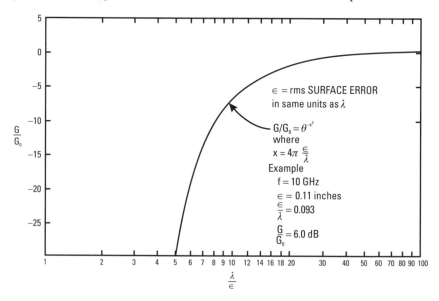

Figure 7.28 Normalized loss due to finite surface error (according to Ruze).

same time. This may be accomplished at a low power level prior to the HPA, or if the requirements dictate, at a high power level prior to the antenna. In the latter case, the outputs of multiple HPAs must be combined in a way to produce minimum RF loss and undesirable interaction among the RF signals. For this to work effectively, the following features must generally be provided:

- Filtering of the carrier after amplification to reduce adjacent channel and out-of-band emissions produced by amplifier nonlinearity;

- Combining of RF carriers on different frequencies for transmission on a common waveguide and antenna feed;

- Low RF loss and acceptable channel distortion (e.g., gain slope and group delay) in providing the above features.

The typical type of device that meets these requirements is an RF multiplexer composed of microwave filters, shown in Figure 7.29. This particular arrangement has five inputs: four enter through individual microwave directional filters, while the fifth enters from the end of the string. The directional filter, shown in Figure 7.30, allows signals outside of the passband to

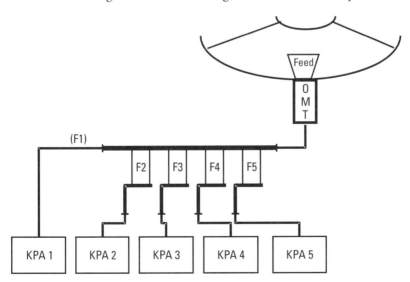

Figure 7.29 A high-power RF multiplexer used to combine the outputs of several klystron power amplifiers in an earth station.

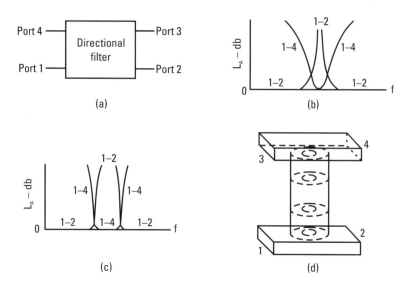

Figure 7.30 Characteristics of an RF directional filter: (a) schematic diagram, (b) transfer characteristic of single cavity, (c) characteristic of multiple cavity, (d) drawing of multiple cavity directional filter.

circulate around the filter, whereas the signals within the passband go through to the other side.

The multiplexer introduces 0.5 to 1.0 dB of loss due to internal resistance and VSWR. For the maximum power of a 3 kW klystron power amplifier (discussed in Section 7.6.1.3), this equates from 300 to 750 watts of excess heat. It becomes an imperative that this heat be removed through conduction and/or radiation so that the equipment does not overheat. Likewise, the combination of all of the outputs, amounting to up to 15 kW in this example, will require high-speed air blowers or even water cooling. Similarly, the small internal dimensions of the antenna feed must withstand the heating, potentially requiring additional antennas for heating reasons alone.

Low-level power combining is perhaps less complex, since a series of hybrids may suffice. There would be approximately 3 dB of loss per hybrid in tandem connection. Another way to determine the total loss is to calculate the division factor, e.g.,

$$L = 10\log(N)$$

where N is the number of inputs to be combined. Thus, there would be 9 dB of direct loss for combining of 8. Additional line loss of about 1 dB would normally be encountered as well. This approach is rather lossy but allows maximum flexibility, since the carriers may be located at any frequencies within the passband of the HPA. The other aspect is that the HPA will be operated under multicarrier conditions and adequate output backoff must be allowed to control intermodulation distortion.

7.6 High-Power Amplification

For earth stations and user terminals that transmit, the high-power amplifier (HPA) is probably the most critical active component. When operating properly, it disappears into the background and tends to be neglected. However, due to the sensitivities involved with DC to RF conversion, amplification, heat generation, and removal, and the need for stable prime power (sometimes at high voltage), HPAs can be challenging when it comes to their selection and operation. We consider here a range of performance from less than one watt to up to 10 kW of output, and both solid-state and vacuum tube configurations. In the past, the main focus was on obtaining high power over the required bandwidth, and DC to RF efficiency was not a significant concern. Introduction of mobile and portable user terminals has made efficiency and durability much more of a concern, although power levels for these applications tend to be below 10 watts.

7.6.1 Amplifier Technology

A typical RF HPA is comprised of four key components: the amplifier module itself, which provides some of the gain and the majority of the RF output power, a low-level driver amplifier to bring the total gain up to meet the overall HPA gain requirement, tuning and bias circuitry required for the particular type of amplifier and application, and a power supply to derive the required voltages and currents from the selected prime power source (e.g., AC for fixed installations and DC for mobile and handheld terminal designs). The primary focus of HPA design is on the power amplifier module itself as it determines the output signal characteristics. In the following sections, we review specific amplifier designs available for earth stations and user terminals.

7.6.1.1 Solid-State Power Amplifier (SSPA)

While the original HPAs in use were based on vacuum tube technology (discussed in the following sections), there has always been a keen interest in using solid-state RF amplifiers for the simple reason that they do not wear out. In addition, there is a perception that they are less complex and hence are more reliable. Whether these benefits are obtained in practice or not depends on many factors because HPA requirements are much different from those of audio amplifiers, computer memory, and digital signal processing circuits. In spite of these issues, there are many options for using solid-state power amplifiers (SSPAs) in earth stations that provide some important advantages in terms of compact design, improved reliability and lifetime, and enhanced linearity.

The first introduction of solid-state RF devices in commercial earth stations was in the LNA and driver amplifier. The SSPA was introduced around 1985 with the advent of the low-cost two-way terminal for low- and medium-speed data transmission. At a power level of around 5 watts, the early SSPA was limited to such thin route applications, although without it, the modern VSAT would not have grown to its current market size. At the time of this writing, SSPAs were on the market reaching up to about 400 watts through paralleling of many stages. This is because the power output of an individual transistor stage is rarely greater than about 20 watts.

The principle output device of the SSPA is the Gallium Arsinide field effect transistor (GaAsFET), which comes from the same family as the GaAsFET used in low noise amplification. In this role, however, device and circuit are different owing to the need to dissipate more heat and for efficiency in converting DC to RF. SSPAs are available at C and Ku bands, and are being developed for Ka band as well; however, the available performance from GaAsFETs definitely decreases as frequency increases. This is why one can obtain SSPAs at C band with output power up to 400 watts (by paralleling stages), while at Ku band, almost nothing is available above 100 watts.

As an example of a high-power SSPA, Figure 7.31 illustrates a unit from CPI with a capability of 200 watts at C band. The power consumption at maximum output is 1,500 watts AC. What is notable about the SSPA is that it does not exhibit the same type of saturation characteristic as a TWTA (to be covered in the next section). In particular, the curve flattens after a point in the transfer characteristic called the 1-dB compression point (defined as the point where the actual characteristic differs from a linear gain, e.g., a straight line, by one dB). It is inadvisable to operate an SSPA into deep saturation, as this can damage GaAsFETs. When operating below the 1-dB compression point, an SSPA offers improved linearity as compared with either a TWTA or KPA. This means that if multicarrier operation is the objective, then it is possible to use an SSPA with a lower saturated power and

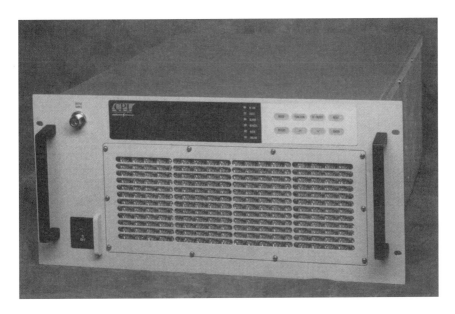

Figure 7.31 C-band solid-state high-power amplifier, capable of producing from 60 to 200 watts (photo courtesy of CPI).

still achieve the C/IM requirements. This can amount to as much as a three-dB difference.

From a Ku- and Ka-band perspective, the prevailing thought is that SSPAs are best suited for low-power single-carrier installations in VSATs and other types of user terminals. A Ku-band transceiver module capable of power up to 25 watts' output is shown in Figure 7.32. An important objective of Ku-band SSPAs is to provide a low-cost device along the lines of the LNBs heavily in use for DTH applications. At the time of this writing, products of this type were beginning to appear on the market, although much effort is expected in bringing the cost in line with consumer applications. The production of a low-cost low-power transmit unit will represent an important breakthrough. Hampering this power amplifier has been the problems of obtaining inexpensive yet reliable GaAsFET devices.

7.6.1.2 Traveling Wave Tube Amplifier (TWTA)

The TWTA has been the traditional HPA for earth stations since the first installations in the INTELSAT system back in the 1960s. Those were 10 kW devices that could operate across the entire 500-MHz uplink range at C band. During the 1970s and 1980s, C- and Ku-band TWTAs in the

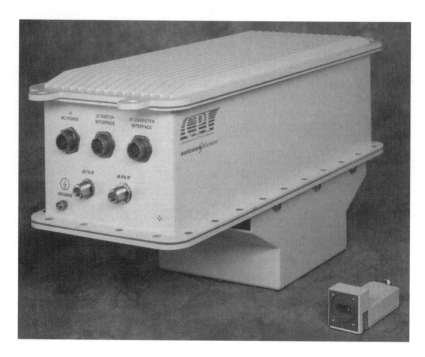

Figure 7.32 Ku-band earth station transceiver package with 6, 16, or 25 watts RF output and 70 MHz IF input (photo courtesy of CPI).

100-to-600-watt range were very popular for commercial earth stations. During the 1990s, TWTAs were introduced at the 800-watt level as C-band and Ku-band amplifiers became common at approximately the same power levels.

Because of their cost and complexity, TWTAs are generally limited to applications in major earth stations and larger user terminals (which could be aptly described as earth stations in their own right). But TWTAs have special features that make them particularly valuable in certain applications:

- A usable RF bandwidth of at least 500 MHz at C through Ka bands;
- Good overall efficiency in converting prime power (AC or DC) into saturated RF output power;
- Ability to handle heat and cold, due to the inherent ruggedness of the traveling wave tube itself;
- Availability of amplifiers in all frequency bands of interest (e.g., L, S, C, X, Ku, and Ka).

This last point on the market for amplifiers is particularly important for cutting-edge applications. For experiments over the Advanced Technology Satellite (ATS) operated by NASA, Ka-band user terminals were provided with TWTAs. A significant issue in using TWTAs is their cost, which can exceed that of the SSPA. However, they will continue to be used below 15 GHz for medium power (e.g., 200 to 1,000 watts) and above 15 GHz for power levels that exceed a few watts.

7.6.1.3 Klystron Power Amplifier (KPA)

Output power requirements that exceed about 1 kW are candidates for the KPA. Historically, KPAs were large, expensive, and temperamental; as a result, their usage was limited to major earth stations used for satellite control and TV transmission. With improvements in amplifier and power supply packaging, KPAs have been brought down to size and, at the same time, become more dependable. An example of one of these compact designs is shown in Figure 7.33. The klystron vacuum tube employs a resonant cavity as part of the amplifier structure and therefore is somewhat limited in bandwidth. This means that the typical KPA will be able to operate over a single

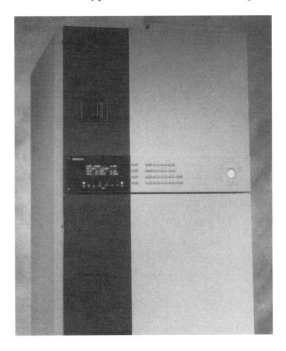

Figure 7.33 C-band 3-kW compact klystron power amplifier (photograph courtesy of CPI).

transponder. The process to change transponders involves mechanical adjustment of the cavity dimensions, which retunes the center frequency accordingly. While tuning is a mechanical action, control is usually extended through electrical actuators.

An example of the block diagram of a 3-kW C-band KPA is provided in Figure 7.34. Some of the features of KPAs include the following:

- Ability to transmit at high power, between 1 and about 4 kW;

- Operation in the higher frequency bands, including C, X, Ku, and Ka bands;

- Considerable commonality among amplifiers in different satellite communication applications (e.g., video, data, satellite control/TT&C, and mobile satellite feeder links).

7.6.2 Application Guidelines

In this section, we provide some technical and operational criteria for comparing the different amplifier types. The following are usually the most important with respect to the application:

- *Maximum power output* (the saturated output power available from the particular amplifier type, delivered for a specified input power);

- *Gain* (both the small signal gain and the saturated gain must be specified, which can be adjusted either locally or remotely using an input attenuator);

- *RF bandwidth* (the minimum usable bandwidth of the amplifier, which must be consistent with the carrier transmission requirement and the possible need to move the carrier within the uplink frequency range—the latter may involve retuning);

- *Linearity* (carrier to intermodulation ratio or digital sideband regrowth);

- *Efficiency* (prime power to RF output);

- *Cost* (purchase price, service, and operation);

- *Size and weight* (transportation and installed);

- *Lifetime and reliability* (device and overall amplifier package);

- *Safety* (installation, operation and maintenance, radiation);

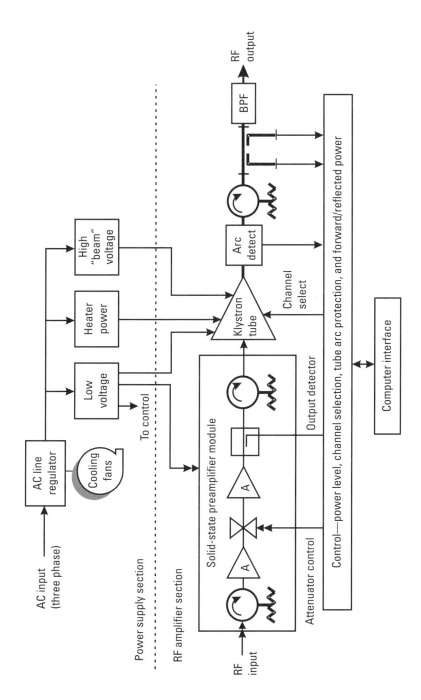

Figure 7.34 Generalized block diagram of a high-power klystron power amplifier.

- *Environmental ruggedness* (resistance to shock, heat, cold, humidity, changes in operating conditions).

7.7 Up- and Down-Conversion

Up- and down-converters provide the translation between intermediate frequency (IF), which is typically 70 MHz or 140 MHz, and the actual uplink and downlink frequencies, respectively. In larger earth stations, the converters are separate units designed for flexibility, ease of maintenance, and stable operation. When performed within a user terminal, the requirements allow low-cost manufacture and integration with the RF or baseband components. Typical block diagrams are provided in Figure 7.35 for a fixed frequency unit that converts between 70 MHz IF and Ku band (e.g., 14.0 to 14.5 GHz uplink and 11.7- to 12.2-GHz downlink). The objective of these converters can be summarized as follows:

- Perform the frequency translation between IF and RF for specific transponders with specified center frequencies and bandwidths;

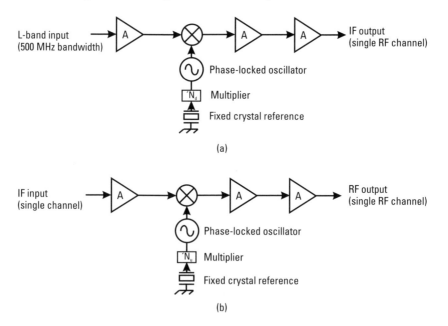

Figure 7.35 Block diagrams of fixed-frequency converters used in earth stations: (a) up converter, (b) down converter.

- Allow for selecting other transponders from time to time, by replacing reference oscillator module or crystal, or with a frequency synthesizer that allows any transponder to be selected (this has become the preferred mode and is more accessible due to technical innovations and reduction of manufacturing cost);

- Minimize the error in translated frequency, through use of stable oscillator design or the provision of an accurate internal or external reference (discussed in Section 8.1.5);

- Provide a satisfactory frequency response in terms of gain flatness, out-of-band rejection, and group delay (much the same as for the satellite transponder itself). This is because the converter is the principle bandwidth-limiting device within the RF portion of the earth station. The analysis of group delay and other impairments is covered in Chapter 8.

Recognized suppliers include Scientific Atlanta, LNR Trexcom (a division of L3 Communications), Miteq, and NEC Corp. Below, we provide the detailed characteristics of one of these products for reference purposes and not as a recommendation.

The LNR Trexcom M2-Series Up and Downconverter, reviewed in Table 7.6, was designed to support services over INTELSAT, EUTELSAT, PanAmSat, and other regional and domestic GEO satellite systems. The units can be used in 1:1 and 1:N redundancy-switched systems, reducing rack space in both fixed earth stations and transportable terminals. Phase noise characteristics are compatible with standard QPSK and eight-phase PSK modulated carriers. The approach taken is that the distortion and error introduced by the upconverter should be substantially below that of the satellite repeater. The unit has adjustment capabilities to permit maintenance personnel to optimize performance for the particular carrier type and transponder characteristic. It may be necessary to make compensating adjustments from time to time, using specialized test equipment such as a frequency counter, microwave link analyzer, and bit error rate test set.

The particular unit is synthesized to permit operation in any transponder and at any frequency within that transponder. Indicated in Table 7.6 is the short-term frequency stability of the oscillator reference, which is a temperature-compensated crystal oscillator. Greater accuracy could be obtained by anchoring the synthesizer to an atomic standard or an external primary standard. Also, phase noise is contributed to the carrier when it passes through the frequency conversion stage. This should be budgeted on

Table 7.6

Detailed Specifications of the LNR Trexcom Low Phase Noise Upconverter

Type	Dual Conversion
Frequency conversion	Positive (No Inversion)
Frequency Selection	(Local or Remote) Synthesizer tuned, 125 KHz steps—(1 KHz steps optional)
Frequency Stability	$\pm 1 \times 10^{-8}$/month, including 10 to 40°C temperature change
First IF Frequency	Above 1 GHz
Reference Frequency	Front Panel Adjust and Monitor
Input Frequency	70 MHz (Option, 140 MHz)
Impendance	50 Ohms
Return loss	23 dB (nom.)
Connector	BNC
Output Frequency	13.75–14.5 GHz (Std.)
Impendance	50 Ohms
Return loss	15 dB (nom.)
Connector	SMA
Level (1 dB compr.)	+6 dBm (min.) (high value option available)
Muting	80 dB (min.)
Transfer Characteristics	
Gain	32 dB (nom), 30 dB (min)
Bandwidth & Ripple	36 MHz, ±0.25 dB (max)
Gain Slope	±0.05 dB/MHz (max.)
Gain Stability	±0.25 dB/day (max)
Group Delay (± 18 MHz)	Linear—±0.05 ns/MHz (max) Parabolic—0.008 ns/MHz (max) Ripple—1.0 ns, p-p (max)
Phase Noise	3 dB better than IESS 308/309
AM to PM Conversion	0.1°/dB (max) for −10 dBm output
Gain Adjust	30 dB in 0.5 dB steps Front Panel Control
Remote Control	RS422/485 (Std.)
Slope Adjustment	±1 dB Front Panel

an end-to-end basis, but is usually only a concern for narrowband signals such as those used in SCPC telephone and low-speed data applications.

References

[1]   Cheung, W. Stephen, and Frederic H. Lovien, *Microwaves Made Simple*, Dedham, MA: Artech House, 1985, p. 44.

[2]   Pritchard, Wilbur, "The calculation of system temperature for a microwave receiver," *The Communications Handbook*, Jerry D. Gibson, ed., Boca Raton, FL: CRC Press, 1997.

[3]   Fujimoto, K., and J. R. Jones, *Mobile Antenna Systems Handbook*, Norwood, MA: Artech House, 1994, p. 465.

[4]   Kitsuregawa, Takashi, *Advanced Technology in Satellite Communication Antennas: Electrical and Mechanical Design,* Norwood, MA: Artech House, 1990.

[5]   Kitsuregawa, *Advanced Technology in Satellite Communication Antennas*, p. 82.

[6]   Elbert, Bruce R., *Introduction to Satellite Communication*, 2d ed., Norwood, MA: Artech House, 1999, p. 283.

[7]   Johnson, Richard C., *Antenna Engineering Handbook*, 3d ed., New York: McGraw-Hill, 1993, pp. 17-21.

[8]   Ruze, J., "Antenna Tolerance Theory: A Review," *IEEE Proc.,* Vol. 54, April 1966, pp. 633–640.

[9]   Chang, Kai, *Handbook of Microwave and Optical Components*, Vol. 1, Microwave & Antenna Components, New York, NY: John Wiley & Sons, 1989, pp.556.

8

Signal Impairments and Analysis Tools

Technical analysis of the ground segment encompasses many fields, ranging from communications engineering to orbital mechanics. Rather than attempt to address every aspect and methodology, we have selected specific approaches that have the most leverage for optimizing technical performance. We first examine some of the more traditional analytical approaches for determining the impairments to modulated signals as they pass through elements in the end-to-end path. Important to this is an understanding of the effects of amplitude and phase nonlinearity, group delay, and intermodulation distortion. The remainder of the chapter reviews three classes of software tools for the PC: SatMaster Pro, a low-cost but effective link budget package; Satellite Tool Kit, a popular orbital visualization tool; and PC-based signal analysis and synthesis packages that can even produce a digital signal processor (DSP) circuit design, if desired. Two that fall into the last category are SystemView, from Elanix, and Software Test Works (SPW), from Cadence Design Systems. This discussion is meant as an introduction and starting point for detailed investigation where required.

8.1   Allocation of Digital Signal Impairments Between Space and Ground

The title of this section suggests that we need to segregate the impairments between those that are produced by the ground segment and those from the space segment. Then, within each segment, we allocate them further to major elements, such as receivers, multiplexers, filters, amplifiers, and in the case of DSPs, to various mathematical operations associated with conversions and manipulations. This allocation process often proves too conservative and so an end-to-end simulation can be beneficial. This is a very time-consuming process, requiring the skills of highly qualified engineers and even scientists. The basic structure for this type of study is presented in Figure 8.1, which indicates how the building-block nature of the end-to-end link impacts service quality. At each block, the communication signal experiences linear and nonlinear forms of distortion, noise and interference, and other impairments that may or may not be immediately apparent.

A list of typical impairments in digital systems is presented in Table 8.1. While most of these impairments will be present in every system, their values and relative meaning will vary greatly from case to case. By assessing, evaluating, allocating, and optimizing these impairments, we gain insight into the leverage points that impact the performance, cost, and even the feasibility of the system. The first five impairments are determined individually for particular elements in the end-to-end transmission system. Some are due to passive devices like filters, while others are introduced by active components within the receivers and HPA. The last row represents the combination of all impairments when acting together on the communication signals. This may not be a simple sum of the individual contributions; hence, it is often necessary to use either hardware or software (computer) simulation to assess the total impact. Some of these tools are discussed at the conclusion of this chapter.

Assignment of these allocations to the ground and space components determines, in large measure, the difficulty of designing and building the particular components. For example, it is much easier to control intermodulation noise on the ground by increasing HPA size than by doing the same in the satellite transponder. Therefore, we would likely allow for more of this type of impairment from the satellite than the earth station. On the other hand, phase noise produced within the satellite would not be correctable on the ground, and so adequate measures must be taken in the design of the space components. Some experience-based guidelines are provided in the following subsections, which relate to each of the rows in Table 8.1.

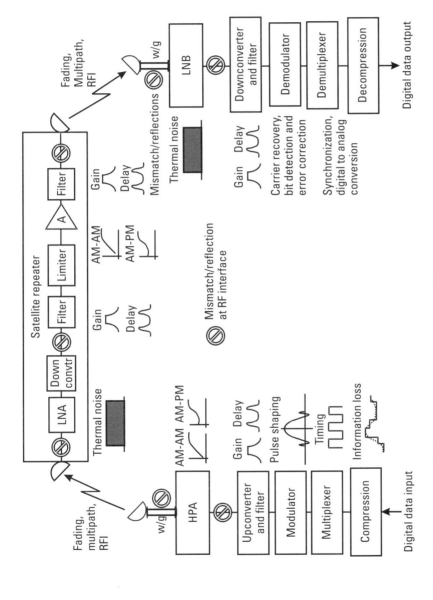

Figure 8.1 Signal impairment contributions in a typical end-to-end satellite link.

Table 8.1
Primary Transmission Impairments in a Satellite Digital Communications Link

Impairment	General impact	Sources (Space or Ground)	Corrective Actions
Intermodulation noise	Added RF noise	Nonlinear amplifiers	Linearizer or proper carrier spacing
Amplitude versus frequency distortion	Variation in carrier levels or modulation crosstalk	Filters and amplifiers	Better tuning, amplitude equalization
Group delay (phase versus frequency distortion)	Alteration of phase modulation, baseband intermodulation, intersymbol interference	Filters	Better design, delay equalization
Amplitude and phase nonlinearity	Intermodulation noise, modulation transfer	Nonlinear amplifiers	Backoff, use of linearizer
Phase noise and jitter	Baseband noise, carrier frequency error	Up- and down-microphonic converters	Improved local oscillators, frequency phase lock control
Combined effects of multiple impairments and nonlinearities	RF and baseband noise, other signal impairments, intersymbol interference	All active elements and filters	Adjustment and optimization of the transmission system design

8.1.1   Intermodulation Noise

All satellite communication systems are designed to overcome the various types of natural and artificial noise that enters through the uplink and downlink. In Section 3.1.1, we described the process of summing the noise to produce the combine C/N for the end-to-end link. We assumed that intermodulation (IM) noise could be considered like thermal noise, which is typically how it is done in the industry. There are circumstances, however, where IM noise characteristics produce a greater impairment to the service than indicated by summing it with the other noise sources. An example of this is coherent IM experienced in FM, where the products contain modulation from the communication signals. As a result, coherent IM can introduce intelligible crosstalk into FM receivers if the product falls within the signal bandwidth.

Narrowband IM products will produce a peaked spectrum that is not flat within the bandwidth of the desired signal. Whether this poses a difficulty for the desired signal or not depends on the transmission parameters of the link. Dynamic IM products caused by external interface sources, alter their frequency and bandwidth, due to the variability of the signals producing them. Such products can sweep in and out of carrier bandwidths, producing intermittent difficulties for normal communication services.

Earth stations that transmit multiple carriers (e.g., SCPC) are potential sources of IM noise. This can be controlled by either using separate transmitters for each carrier (which is an expensive solution) or by increasing the saturated power of the HPA to allow greater multicarrier backoff. Major earth stations can be equipped with TWTA and klystron power amplifier (KPA) systems with the needed extra power. With adequate output backoff (e.g., 3 dB or greater), IM noise can be held to 20 dB below the carrier. Some ground segments employ larger VSAT antenna that have the capability to transmit two or more carriers. This is a way to combine voice and data within the same system. The HPA in this case would need to be chosen with a total system tradeoff in mind, e.g., considering the intermodulation noise of the VSAT combined with that of the satellite transponder.

DBS satellite systems are uniformly operated on a single-carrier-per-transponder basis so that maximum signal power and minimum IM noise are provided on a continuous basis. Multicarrier services like SCPC telephony, VSAT networks, and video teleconferencing using FDMA will produce very significant amounts of IM noise within a given transponder. In a typical link budget, the carrier-to-intermodulation-noise ratio (C/IM) will have a value comparable to the downlink C/N, which in and of itself will decrease their combination by 3 dB relative to the downlink C/N alone. An example of this trade can be seen in Figure 8.2. Increasing the saturated power is not feasible once the satellite has been designed, constructed, and launched (an exception is for repeaters with the ability to parallel two amplifiers or increase saturated power within the amplifier, both by ground command).

A recent innovation in onboard processing called Skyplex can greatly enhance the operational flexibility and system capacity of point-to-multipoint information distribution [1]. The concept behind Skyplex is to employ SCPC digital uplinks that are demodulated and combined in an onboard multiplexer for transmission to the ground on an MCPC basis. This permits several individual uplinks to originate material from different locations and use the satellite as the point of aggregation (instead of a common uplink broadcast center). Some of the characteristics of this application are presented in Table 8.2. By combining the streams into one broadband downlink, Skyplex prevents multicarrier operation of the TWT. A performance improvement of between 3 and 6 dB of output power is the result (or

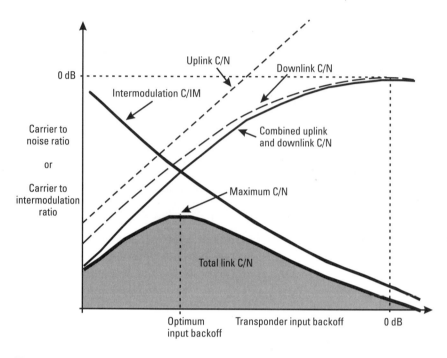

Figure 8.2 Tradeoff between C/N and C/IM in an FDMA transponder or HPA.

Table 8.2
Summary of Skyplex Characteristics in Uplink Direction

Number of uplink carriers per transponder	Up to 8
Uplink bit rate	From 2 to 7 Mbps
Downlink bit rate	55 Mbps
Modulation	QPSK
Access type	SCPC or TDMA
Carrier frequency stability	±108 kHz
Operating E_b/N_0	10.6 dB
Data format	DVB and MPEG-2

equivalently, an increase in transponder capacity of a factor of two to four).
Skyplex went into operation on the EUTELSAT Hot Bird satellites in 1999
and can be introduced in other regions that employ the DVB standard.

8.1.2   Amplitude Versus Frequency Distortion

This particular impairment is common to all communication channels with bandpass filters, as found in the transmitting earth station, transponder, and receiving earth station. Ideally, the filter should be flat as a board within the usable bandwidth, and have very steep out-of-band skirts to prevent adjacent channel interference. Out-of-band rejection that is not ideal will allow some sideband and IM energy to pass into adjacent channels. It is a common property that filters with better out-of-band rejection have poorer inband properties, but performance of modern microwave and digital baseband filters is substantially better than what was possible even in the early 1990s. This is due to improved analytical and manufacturing techniques for hardware filters and greater use of the ideal response properties of digital filters within the DSP.

The three basic filter properties that produce amplitude distortion are identified in Figure 8.3. These are as follows:

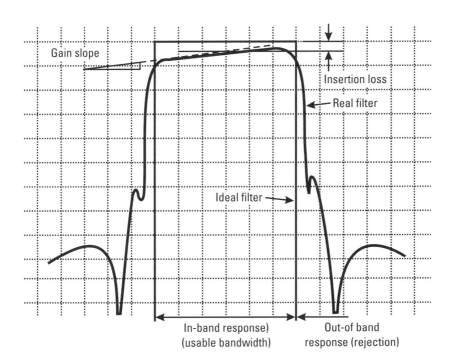

Figure 8.3 Ideal and real filter amplitude responses (gain versus frequency).

1. In-band ripple, which has a peak-to-peak variation produced by the number of tuned circuits and their coupling characteristics. Wideband communication signals that have bandwidth much greater than that of one ripple cycle are not affected as long as the ripple variation is less than about 0.25 dB. The effect of greater ripple amplitude can only be assessed by analysis or test.

2. Gain slope, which is a steady tilt of the frequency response, measured in dB/MHz, with either a positive or negative slope. This impairment interacts with frequency modulation on a carrier, if present, and produces amplitude modulation with the same content; it may also cause I-Q coupling in QPSK.

3. Band-edge droop, which is a rounding of the amplitude response, giving the appearance of a frown (convex downward curvature), as shown in Figure 8.3. Impairment to a centered carrier is usually minimum; however, narrowband carriers that are off center will experience some degree of gain slope.

In practice, amplitude versus frequency distortion is usually expressed as a linear combination (in dB) of the three distortion types. An example is as follows:

$$X \text{ dB per MHz; plus } Y \text{ dB per MHz}^2; \text{ plus } Z \text{ dB per MHz}^3$$

The effect on signals that traverse the channel may or may not also be a linear sum. In some cases, adding the effects produces greater impairment than is experienced by the effects simultaneously. All three components of amplitude response (x, y, and z) may be partially compensated in the transmitting and receiving earth station.

8.1.3   Group Delay Distortion

The ideal communications channel is one with a flat amplitude versus frequency response and constant time delay over the same bandwidth. Since time delay is the first derivative of phase, we can therefore conclude that the ideal is for a linear slope of phase versus frequency. Any deviation from constant delay versus frequency introduces an impairment to carriers that pass through the channel. We term this *group delay distortion* (the reference to "group" means that a modulated signal is used to measure the delay over the bandwidth of a range of frequencies rather than a single frequency) [2]. It is possible to compensate for nonlinear phase using phase through delay equalization, which is common in data communications where phase distortion is a concern.

Group delay is measured in microseconds (ms) or nanoseconds (ns), and is a relative measure. An example of such a characteristic is shown in Figure 8.4 for a typical transponder channel over a C-band satellite. The group delay is typically minimum at the center of the channel, where we arbitrarily set its value to zero. It then increases toward the edge of the transponder, hits a maximum somewhere out of band, then goes either downward or upward, depending on the filter transfer characteristics. While we assume that the group delay is zero at the center, there is in fact a fixed delay associated with passage of the signal through the filter.

Group delay can be a serious impairment for FM and PSK. In digital systems, the effect on PSK transmission is to distort the transitions between symbols. The resulting received pulse train deviates from the ideal shaping that was produced by the modulator. A way around the problem is to use delay equalization on the same side of any nonlinear amplifier in the channel. For example, group delay in the transmitting earth station can be compensated in the same station or onboard the satellite ahead of any nonlinear devices. In the same manner, group delay after the satellite power amplifier would be compensated either on the downlink side of the satellite repeater or within the receiving earth station. One issue that has to be dealt with is the number of equalizers that have to be included. For example, equalizing

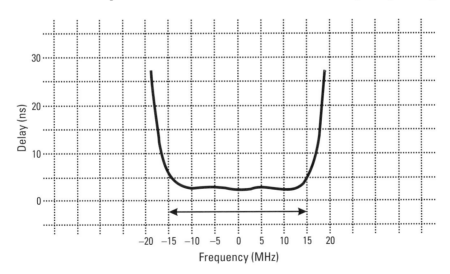

Figure 8.4 Typical group delay characteristic (delay versus frequency).

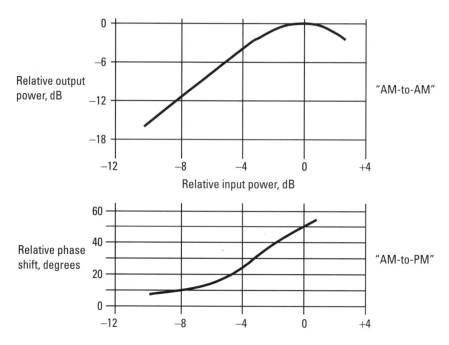

Figure 8.5 TWT nonlinear characteristics (AM-to-AM and AM-to-PM).

the transmitting earth station is fairly simple since we can compensate for a fixed level of impairment. Doing this within the receiving earth station may require as many equalizers as there are receivers, which for a DTH system could number in the millions. If this can be done using digital signal processing, the equalization may be trivial to include in all receivers.

8.1.4   Amplitude and Phase Nonlinearity

Passive devices like transmission lines, reflector antennas, and filters exhibit linear types of impairments like gain slope and parabolic group delay distortion. Active devices used to amplify, translate in frequency, and modify the signal structure involve some form of nonlinearity that is sensitive to the signal amplitude (or power level) yet independent of frequency. The classic TWT characteristic, shown in Figure 8.5, is a dominant source of amplitude and phase nonlinearity for satellite communication services. However, readers should be aware that any active device will behave in a similar way. Transistor amplifiers, like GaAsFETs, have generally less distortion versus input power. In either case, it is possible to compensate by introducing distortion that is opposite-going to either the amplitude or phase or both. Of course, linearization cannot cause the output power to be greater than the amplifier is capable of delivering.

Uncompensated amplitude or phase nonlinearity may have a significant impact on communications service quality. The primary effects are as follows:

- *RF intermodulation distortion*, covered earlier in this chapter. If the driving signals are much too strong, then it is possible for the active device to generate a wide spectrum of unwanted intermodulation products, affecting services in adjacent bands.

- *AM-to-PM transfer*, which is a mechanism for transferring AM from one signal to PM on the same or, in the case of multicarrier operation, to other signals passing through the same amplifier. It is a case of the strongest carrier transferring its AM (if present) onto weaker carriers that share the bandwidth of the channel. The AM-to-PM transfer coefficient is the first derivative of phase shift versus power input, measured in degrees/dB.

Amplitude limiting is an effect on strong carriers that drive the amplifier or other active device toward its point of maximum output (e.g., the saturation point). At low levels of drive, most active devices are nearly linear (constant gain). Moving toward saturation, the signal experiences some degree of limiting, causing flat-topping of the time waveform. In the worst case, the device acts as an ideal limiter, producing a pulse stream at the carrier frequency and harmonics in the output.

8.1.5 Frequency Stability and Phase Noise

Modulators, frequency converters, and demodulators contain local oscillators that allow the carrier to be translated either up or down in frequency. A real oscillator cannot generate a pure sine wave: it will add to the variability of the signals as well as the baseband noise that is reproduced at the receiving end of the link. For a typical mixer stage, the various forms of noise on the local oscillator are transferred directly to the carrier. Providing adequate cleanliness for the LO is therefore a critical aspect of the design of the ground segment and the satellite as well.

Any microwave frequency source can be stabilized by phase locking its frequency to a low-drift, temperature-stabilized crystal reference [3]. Further improvement is obtained by using an atomic frequency standard as the reference. However, the short-term stability of the source is not as easily controlled, and constitutes a principal source of error or degraded performance in many microwave systems. The rather complex methods by which short-term stability must be measured cause us to express the measurement results in unusual terms, which are subject to misinterpretation and confusion.

The nature of short-term instability is that it is a random phenomenon—small, relatively slow, nonsystematic changes in the long-term-average frequency (the "carrier" frequency). One way of thinking about this random variation is to consider that it is the result of frequency-modulating the carrier by a complex signal having a limited "spectrum signature"; in other words, the spectral content of the modulating signal is not infinite in bandwidth with the same amplitude at all frequencies, like that of "white noise," but has most of its energy concentrated at low frequencies, with the amplitude of its frequency components falling off rapidly at higher frequencies. Figure 8.6 shows the spectrum of a signal that would, if frequency-modulated a carrier, result in typical short-term frequency instability. Figure 8.7 shows the actual spectral distribution of phase noise after demodulation from a carrier.

If we recall that there is a physical correspondence between frequency modulation and phase modulation (i.e., that frequency deviation is equal to the rate of change of phase deviation), then it is possible to express a particular short-term instability in terms of either "FM noise" or "phase noise." As we shall see, the measurement of phase noise, either above or below the

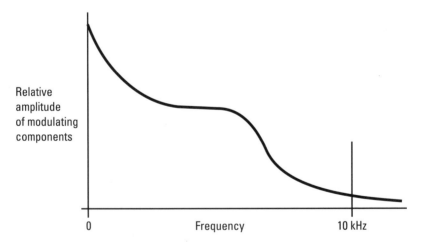

Figure 8.6 Spectrum of a signal that would, if frequency-modulating a carrier, result in typical short-term frequency instability.

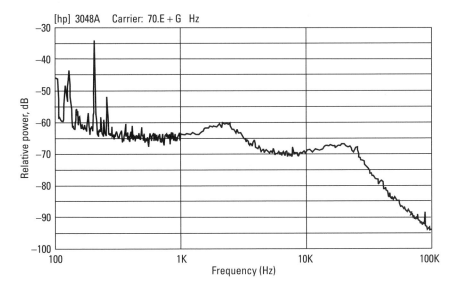

Figure 8.7 Phase noise sideband profile for a downconverted continuous wave tone generated by a 14.5-GHz VSAT terminal.

carrier (single-sideband phase noise) is preferable to measuring frequency deviations when the deviations are small. Another way of thinking about the short-term instability of the signal source is simply to say that the small, random frequency changes cause "phase jitter" in the carrier. For some very critical applications, the measurement and interpretation of phase jitter is the only practical way of evaluating the short-term stability of a signal source.

All of the measurements and calculations discussed in the foregoing text have assumed that the random phenomenon we call short-term frequency or phase change can be averaged over a bandwidth, weighted, and expressed as an equivalent time-invariant quantity. In fact, instantaneous values of random functions may be very much larger than such averages, and, in microwave systems requiring the highest order of stability, even these very brief and/or very-seldom-occurring excursions are significant.

Another source of frequency error is Doppler shift, governed by the formula

$$\Delta f = V_r f / c$$

where V_r is the range rate in meters per second, f is the frequency of the transmitted wave, and c is the speed of light (e.g., 2.98×10^8 meters per second in vacuum). For example, a relative velocity of the earth station with respect to the satellite of 100 m/s at 12 GHz would produce a Doppler shift of 4,027 kHz (4.027 Hz/KHz). This represents an increase in frequency if earth station and satellite are moving toward each other, and a decrease if away. Let us say that our carrier has a bandwidth of 10 MHz, then this shift represents only 0.04 % of frequency error and can be ignored. If, on the other hand, the carrier bandwidth is much narrower, say 10 KHz, then the error is substantially greater in relative terms (e.g., 40%).

Frequency stability is the ratio obtained by taking the absolute frequency error (the worst-case value of the difference between the actual frequency and the assigned frequency) and dividing it by the assigned frequency. Examples of types of oscillators and their typical values of frequency stability are provided in Table 8.3.

The stability of microwave oscillators such as a synthesizer, DRO, or YIG oscillator, can be improved through the use of a phase lock loop (PLL). This can produce an excellent blending of good short-term stability from an oscillator device with excellent long-term stability available from an atomic standard or GPS. As shown in Figure 8.8, the voltage controlled oscillator (VCO) output is divided down by an integer and compared with that of

Table 8.3
Typical Types of Local Oscillators and Their Values of Frequency Stability

Type of Oscillator	Assigned Frequency	Frequency Stability, per month	Frequency Stability, over life
Quartz crystal (not temperature stabilized)	100 MHz	10^{-3}	10^{-6}
Temperature-stabilized quartz crystal	100 MHz	10^{-7}	10^{-6}
Frequency synthesizer	By design	Based on reference	Based on reference
Dielectric resonator oscillator (DRO)	1 to 14 GHz	10^{-5} (based on crystal reference)	10^{-6}
Yttrium iron-garnet (YIG) oscillator	1 to 100 GHz	DC current controlled	
Rubidium atomic standard	—	10^{-12}	10^{-11}
Cesium atomic standard	—	10^{-13}	10^{-13}
Hydrogen atomic standard	—	10^{-15}	10^{-15}
Corrected quartz oscillator (using GPS)	—	10^{-12}	10^{-13}

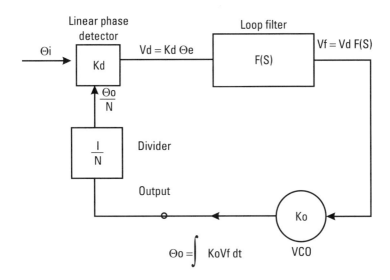

Figure 8.8 Basic phase-lock loop feedback system.

a highly stable crystal reference [4]. The difference is filtered and detected before applying it as a correction input to the VCO. When first turned on, the VCO frequency will be pulled into the reference, requiring a settling time.

8.2 Software Design Tools

One of the benefits of the PC is its ability to perform mathematical analyses and provide excellent visual displays for typical satellite communications problems. For most situations, this is a huge improvement over the scientific calculator, simple spreadsheets, and older (and obsolete) mainframe software, much of which was custom written in years past. We review in this section three PC-based software packages that allow ground segment and earth station engineers to do their jobs more effectively and accurately. Reviewed below are Satellite Tool Kit, an orbit and coverage visualization product; Sat-Master Pro, an effective link budget generation and analysis package; and System View and SPW, two sophisticated signal and system analysis tools that can help predict end-to-end performance. The discussion that follows is not exhaustive, as it would take several volumes to represent this software as well as other equally valuable tools for the PC.

8.2.1   Orbit Visualization (STK)

The Satellite Tool Kit (STK)® is an off-the-shelf PC software computer-aided engineering (CAE) package for orbit analysis, determining earth station pointing angles, and making satellite coverage predictions. This graphical-oriented tool offers individual stand-along capability and cross-discipline understanding to analysts and system engineers at many levels. STK currently supports preliminary definition of requirements, design trade-offs, comparison of orbit and launch, and preliminary evaluation of operating strategies for space and ground segments.

The package is available as a free download from www.stk.com. in the basic configuration with two-dimension visualization. When combined with other modules that are purchased from Analytical Graphics, Inc. (AGI), the tool provides many integrated functions: defining and propagating satellite orbits, determining visibility areas and times, computing antenna and sensor pointing angles, and generating results in textual and color graphical formats.

Based on simple user inputs, STK generates paths for a variety of space- and ground-based objects, such as satellites, ships, aircraft, and land vehicles. An example of an orbit visualization for a LEO constellation is provided in Figure 8.9. STK provides animation capabilities and a two-dimensional map background for visualizing the path of these vehicles over time. STK can

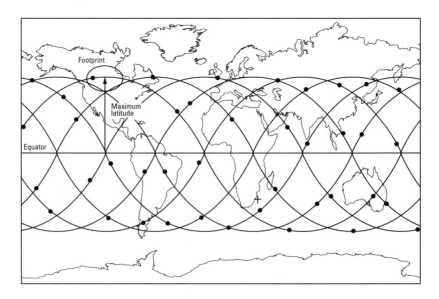

Figure 8.9 Ground track for a LEO constellation as presented by Satellite Tool Kit (image courtesy of Analytical Graphics, Inc.).

model the field of view available to these objects, enabling the user to determine whether the object can "see" its target (e.g., earth station locations or satellites). Along with these displays, STK computes and outputs access times, position, and other key information.

STK's browser window allows the user to add/remove objects, define object properties, and manipulate objects to analyze relationships or summarize key information. Individual objects can be defined by satellite orbit type; graphic characteristics, such as azimuth-elevation masks for earth stations and other targets; constraints, such as sensor ground sample distance; and special properties, such as model attributes.

The object-oriented approach recognizes 14 classes of objects that can be manipulated by the user, including: application (the highest-level object definition, corresponding to the simulation model itself), ground vehicle, area target, scenario, missile, planet, satellite, launch vehicle, star, aircraft, facility, sensor, ship, and target. As objects are created and defined, they compose a scenario—a depiction of the events and relationships between space- and ground-based objects for a particular time period.

The background provides the user with a geographic frame of reference for each object contained in the scenario. The user may select from 10 map projections, including: azimuth equidistant, Mercator, perspective, equidistant cylindrical, Miller, sinusoidal, Hammer-Aitoff, orthographic, stereographic, and Mollweide. STK can determine instances when one object can access or "see" another object and display the corresponding access graphics in the map window. Simulation results display line of sight, sensor field of view, object constraint settings, and violations of constraints. Multiple map windows can be displayed, each employing different map colors, geographic features, display formats, and latitude/longitude line spacings.

With additional software modules purchased separately, the user can apply STK 4.0's 3-D animation capabilities to understand time-based relationships and interactions between objects. Animations can depict satellite ground tracks during each pass, or the intersections of sensors and the Earth. The animation toolbar allows the user to select start/end time, time step, and animation speed. As the animation plays, the user can zoom in or out. In addition, individual screens may be printed for presentation purposes.

AGI maintains an up-to-date database of over 8,000 cataloged objects at www.stk.com. Data include two-line element sets (TLE's), Space Surveillance Catalog (SSC) numbers, common names, launch dates, launch times, apogee, perigee, activity state, and more. This database is particularly useful for inserting objects into STK as vehicles for analysis, as the TLE in the database propagates the vehicle's orbit automatically. Updated three times each week, the database is used by the STK Close Approach (CAT), which references objects in the database to obtain an estimate of approach to an existing vehicle.

8.2.2 Detailed Link Budget (SatMaster)

Link budgets are essential to satellite system design and earth station specification. While they are based on a set of well-understood relationships, they have been very individualized, according to the engineer's knowledge and approach to the problem. Many have tried to standardize link budgets, but there continues to be reluctance by experienced satellite communications engineers to relinquish their individual style and practice. While much can be said for maintaining tight control over the detailed engineering approach that we each take, the variation in formats has complicated the matter of transferring knowledge and even results of link budget analyses, because engineers often have difficulty understanding the assumptions used by others. While we do not recommend a particular off-the-shelf software package for detailed engineering and design, there still is benefit in some of these tools for common problems involving GEO satellite links in the FSS and BSS frequency bands. One relatively inexpensive yet complete tool is SatMaster Pro, available over the Internet.

Provided by Arrowe Technical Services in the United Kingdom (www.arrowe.co.uk), SatMaster, version Mk 5.4, is a reasonably versatile tool for standardized analog and digital link budgets. There are separate link budget formats for analog and digital systems, providing individual uplink and downlink, as well as combined versions. Calculators are included for antenna aiming, dual/multi feed positioning, TVRO dish sizing, and polar mount alignment. Files of simplified EIRP, SFD (saturation flux density), and G/T maps are included for some satellites. Users can access tabular and graphical data of magnetic variation, solar transit times, off-axis gains, rain attenuation, and atmospheric absorption according to recognized models (e.g., ITU-R and Crane). An example of a complete Ku-band digital video link budget is provided in Table 8.4. System requirements for SatMaster include: Windows 95, 98, or NT4.0 (Intel), IBM 486 or better with 8MByte RAM (Win9x) or 16MByte (NT4.0), and 10MByte free hard disk space.

The calculator can determine HPA size, uplink power requirements, bandwidth, and power usage per carrier. Predictions are made of atmospheric losses and rain fade margins based on availability on uplink and downlink. The link may use information rates from 9 Kbps to over 40 Mbps. Output backoff and transponder intermodulation interference may be

Table 8.4

Sample of Link Budget Output Produced by SatMaster Pro

Digital Downlink (.45m Antenna)
Monday, May 01, 2000

Site name	Los Angeles
Satellite name	DIRECTV 1R

Input Parameters	Value	Units
Site latitude	34.00N	degrees
Site longitude	118.28W	degrees
Magnetic variation	13.9E	degrees
Site altitude	0	km
Spot beam polarization cant	0	degrees
Frequency	12.2	GHz
Antenna aperture	.45	meters
Antenna efficiency/gain	60	% (* prefix dBi)
Antenna noise	50	K
LNB noise figure/temp	*80	dB (* prefix K)
Coupling loss (waveguides, polarizers)	.25	dB
Antenna loss (aging, pointing, polarization)	0	dB
LNB gain	50	dB
LNB load impedance	50	Ohms
Co-channel interference C/I	30	dB
Adjacent satellite interference C/I	15	dB
Signal availability (average year)	99.5	%
Rain-climatic zone (* prefix Crane)	E	
Water vapor density	10	g/m3
Surface temperature	20	degrees C
Polarization	Circular	
Satellite longitude	101.20W	degrees
EIRP	50	dBW
RF/IF bandwidth	27	MHz
Information Rate	30	Mbps
Overhead	10	%
FEC Code Rate	.75	

Table 8.4 (continued)

Input Parameters	Value	Units
Spread Factor	1.2	
Threshold IRD Eb/No	6	dB
Modulation	4-PSK	

Satellite Look Angles	Value	Units
Elevation	46.43	degrees
True azimuth	151.21	degrees
Azimuth compass bearing	137.31	degrees

Modified Polar Mount Settings	Value	Units
Polar axis	34.65	degrees
Polar elevation	55.35	degrees
Declination offset	4.87	degrees
Apex declination	39.52	degrees
Apex elevation	50.48	degrees

Clear/Degraded Sky Results	Clear	Rain	Units
Path distance to satellite	37,310.98	37,310.98	km
Free space path loss	205.61	205.61	dB
Spreading loss	162.43	162.43	dB/m2
Atmospheric gaseous absorption	0.11	0.11	dB
Tropospheric scintillation fading	0.21	0.21	dB
Total atmospheric losses	0.32	0.32	dB
Effective area of antenna	−10.20	−10.20	dB/m2
Antenna gain	32.98	32.98	dBi
LNB noise temperature	80.00	80.00	K
Signal availability (worst month)	100.000	98.440	%
Attenuation due to precipitation	0.00	0.64	dB
Noise increase due to precipitation	0.00	0.91	dB
System noise temperature	143.43	176.92	K

Table 8.4 (continued)

Clear/Degraded Sky Results	Clear	Rain	Units
Downlink degradation (DND)	0.00	1.55	dB
Cross polar discrimination (XPD)		37.17	dB
Power flux density (PFD)	−112.74	−113.38	dBW/m2
Carrier power at LNB output	−73.20	−73.83	dBW
Carrier level at LNB output (50 Ohm)	63.79	63.15	dBuV
Carrier level at LNB output (50 Ohm)	3.79	3.15	dBmV
Figure of merit (G/T)	11.16	10.25	dB/K
C/No thermal	83.84	82.29	dB.Hz
C/Io co-channel interference	104.22	104.22	dB.Hz
C/Io adjacent satellite interference	89.22	89.22	dB.Hz
C/(No*Io)	82.70	81.46	dB.Hz
C/(N*I)	8.48	7.25	dB
Eb/(No*Io)	7.51	6.28	dB
Threshold IRD Eb/No	6	6	dB
Margin	1.51	0.28	dB

Additional Data	Value	Units
Information rate (inc overhead)	33.0000	Mbps
Transmit rate	44.0000	Mbps
Symbol Rate	22.0000	MBaud
Occupied bandwidth	26.4000	MHz

estimated and the budgets may employ BPSK, QPSK, or M-SK with various FEC code rates. The software is intended for VSAT networks, broadcast TV, SNG, Ku trucks, data transmission, and radio station feeds.

8.2.3 Signal Simulation and Communication Analysis Software Tools

There are several software packages on the market that allow simplified analyses and simulation of a communication system. Two of the leaders are SystemView, offered by ELANIX of Westlake Village, California (www.elanix.com), and Cierto Signal Processing WorkSystem (SPW), offered by Cadence Design Systems (www.cadence.com). On the low-cost end of the spectrum, Artech House offers VisSim/microicom simulation and design software (www.artechhouse.com). Another option is to build the simulation with Matlab, a versatile but challenging system analysis tool. A common characteristic is to allow the user to build a model of the end-to-end transmission system, including modem, frequency conversion, amplification, filtering, and various forms of noise and interference. The actual analysis and prediction of performance (e.g., BER and eye opening) is done through simulation and FFT techniques. Brief descriptions of both of these packages are provided below. However, readers need to be aware that each is a very powerful tool that also represents a considerable learning investment to grasp the theory, methodology, and user interface.

8.2.3.1   SystemView

SystemView by ELANIX is an example of a high-level conceptual design and evaluation engine that has an effective graphical user interface and real-time analysis environment. Running under Windows NT, Windows 95, and Windows 98, SystemView may be used to design, develop, and test end-to-end communications systems and their components. It has been applied to analysis problems in wireless communications (GSM, CDMA, and TDMA), modem design (QPSK, QAM, and FSK), spread spectrum, audio processing, radar, sonar, signal analysis, signal intelligence, digital beamforming, and direction finding.

The methodologies and facilities within SystemView include:

- True dynamic system simulator;

- Intuitive sampled data (z-domain) and continuous (Laplace domain) system specification;

- Multirate systems, multiple parallel systems;

- Mixed time-continuous and time-discrete systems;

- Graphical finite impulse response (FIR) filter design, including low-pass, bandpass, highpass, band-reject, Hilbert (90 degree phase shift), and differentiators;

- Extensive library of infinite impulse response (IIR) filters including multipole Bessel, Butterworth, Chebyshev, and Linear Phase;

- Fast Fourier transform (FFT) types: magnitude, squared magnitude, spectrum analyzer, power spectral density, phase, and group delay;

- A complement of logic functions, switches, and nonlinear devices;

- Libraries of sources, sinks, functions, operators, and metasystems;

- External and internal signal sources and sinks;

- Built-in system diagnostics and connection checks.

A typical end-to-end system model and the associated time waveforms using SystemView are presented in Figure 8.10. This is only a very simple example of what can be put together and what can be learned. As in the case of STK, there is a considerable learning curve to SystemView, but ELANIX does cooperate by providing free training and a software demonstration. The product is flexible in that once the model is correctly built (using a simple GUI drawing toolbox), it is a simple matter to run multiple simulations. Each time, the program gathers statistics and makes plots available for detailed review. An added feature is that the software may be used to design circuits and chips through an interface to the appropriate CAD/CAM products. In this way, the simulation may be transformed into actual hardware. The informative ELANIX Web page and helpful staff make learning and using this product a positive experience.

8.2.3.2 Cielo Signal Processing WorkSystem (SPW)

Developed originally as a tool to simulate complex analog/digital communications links, SPW is now a full suite of tools and methodologies that take the designer from top-level architecture to subsystem and down to the electronic component level. This is facilitated by the integration of SPW with the suite of CAE tools provided by Cadence Design Systems. Along with its libraries, SPW modules and graphical tools permit the analyst to diagram the system onscreen in an executable form. The block diagram becomes a specification for the operation of the system and circuit. One can develop a virtual prototype for a wide range of communications transmission and processing systems, integrating analog and digital in the same simulation. Under simulation, the signal flow block diagram performs closely to the actual system. Data is collected anywhere a probe is placed in the diagram, and SPW analyzes the data and displays the outcome in a variety of numerical and graphical formats. Simulations using SPW have proven effective in predicting the end-to-end performance of DVB and MPEG-2 transmissions through earth

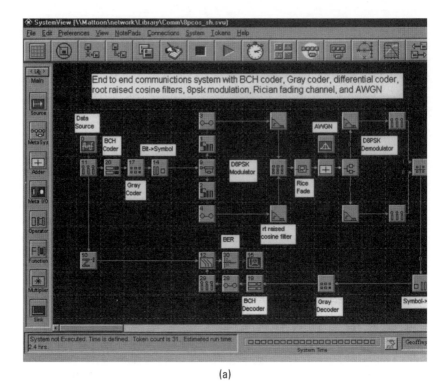

(a)

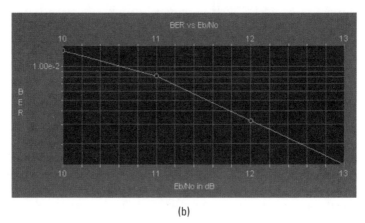

(b)

Figure 8.10 The System View System Window, illustrating (a) a system block diagram and (b) output waveform (image courtesy of Elanix).

stations and satellite repeater under various impairments including group delay and AM-to-PM distortion.

The major components of SPW include:

- *Block Diagram Editor.* This allows the analyst to specify the system by placing and interconnecting blocks, each of which represent a piece of functionality. These may represent a single function or an entire multilevel system, and may come from the library or be developed as part of the particular project. Examples of blocks include modulators, demodulators, nonlinear amplifiers, DSPs, timing circuits, and filters.

- *Signal Calculator.* This is used to import, create, edit, and analyze signals. Basic waveforms and noise signals can be chosen from push buttons, and mathematical operations like sampling may be performed on them. Signals can be cut and pasted within a window and between windows, combined with other signals, put through linear or nonlinear filters, and examined in a variety of ways. The latter include auto and cross-correlation, eye diagrams, histograms, X versus Y plots, and FFTs.

- *Convergence Simulation Architecture.* This enables effective management of mixed-level designs and mixed data types across various design layers. This capability is used when designing a complex system whose overall capability derives more from the interaction of the subcomponents than from their individual functionality.

Cierto SPW is written in object-oriented C++ to enable the analyst to capture the full system end to end while representing the individual components using different levels of abstraction. It can be used to design a component, such as a receiver "system on a chip" (SoC) and evaluate its performance in an end-to-end satellite communications transmission system with earth station and satellite repeater active and passive elements [5]. Further, the Cielo hardware design system (HDS) can then specify the needed digital logic and software code.

References

[1]   Carducci, F., and R. Novello, "On-board Multiplexing: The Skyplex Transponder and the Multimedia Highways," 17th AIAA International Communications Satellite Systems Conference and Exhibit, a collection of technical papers, February 23–27, 1998, Yokohama, Japan, American Institute of Aeronautics and Astronautics, New York, 1998, p. 461.

[2]   Lance, Algie L., "Microwave Measurements," *Handbook of Microwave and Optical Components,* Vol. 1, *Microwave Passive & Antenna Components,* ed. Kai Chang, New York: Wiley, 1989, p. 495.

[3]   "Three Techniques for Measuring and Interpreting Short-Term Frequency Stability," Communication Techniques Incorporated, http://www.cti-inc.com/products/app_notes6htm.

[4]   Cheah, Jonathon Y. C., *Practical Wireless Data Modem Design,* Norwood, MA: Artech House, 1999, p. 148.

[5]   Chang, Henry, et al., *Surviving the SoC Revolution: A Guide to Platform Based Design,* Boston: Kluwer, 1999.

9

Fixed and Mobile User Terminals

A user terminal is a satellite communications earth station designed to be used and operated by the end user. Examples of end-user situations, reviewed in Table 9.1, include cable TV systems and TV stations, branch offices of companies, military units, schools, aircraft and vehicles, and individual subscribers. Unit costs are usually held down because of the potentially large quantities involved (numbering in the hundreds of thousands to millions for a particular model) and a requirement for affordability by end users who may make the financial commitment themselves. In fact, large industry segments like DTH TV and MSS telephony can only exist and prosper if user-terminal costs are within reach of mass markets.

The design of a user terminal is usually highly tied to the particular requirements of the intended service and end user. For example, a DTH receive-only TV terminal is able to receive the wideband Ku-band digital downlink of a GEO satellite, demodulate and process the data stream, and produce video and audio of the desired quality. It incorporates all of the technical features of a receiving earth station as well as those of a modern digital TV set-top box. A VSAT, on the other hand, is an interactive data and, if required, voice earth station within a package about the size of a PC. Unlike the closed architecture of the DTH receiver, a VSAT can be configured for service through selection of appropriate modules and software. Therefore, VSAT terminals behave more like traditional earth stations, although their cost is substantially lower to justify their application in corporate branch

Table 9.1
User Terminals Used in Satellite Communications, GEO, and Non-GEO Systems

Application (Satellite System)	Terminal Type	RF Terminal	Baseband and User Interface
Cable and broadcast TV (GEO FSS bent pipe, C or Ku band)	Fixed earth station at cable head end or TV station	Receive-only, fixed dish (2 to 10m); outdoor electronics; transmit may be included	Analog or digital receivers (one per channel) to recover baseband; includes network control facilities
Direct-to-home TV broadcasting (GEO BSS bent pipe, Ku band)	Individual fixed earth station at home	Receive-only, fixed dish (45 to 90 cm); outdoor electronics	Digital set-top box to recover baseband and deliver programming through friendly user interface
Private business networks, low to medium date rate, voice, TVRO (GEO FSS bent pipe, C or Ku band)	VSAT installed at remote locations	Transmit and receive, fixed dish (1.2 to 2.8m); outdoor electronics with SSPA module	Digital indoor unit containing modem, multiple access controller, and interface port cards; includes network management facilities
Shipboard voice/data communications and distress (GEO MSS bent pipe, L band)	Inmarsat maritime terminal	Transmit and receive dish (1m) or omni antenna; outdoor electronics with SSPA module	Digital indoor unit containing modem, multiple access controller, and user interface
Mobile and transportable (including "fly away") terminals for remote TV origination (GEO FSS bent pipe, Ku band)	Mobile van or trailer containing equipment and operating console	Transmit and receive dish (1.8 to 3m); electronics either inside van or outside in close proximity, TWTA	Rack of equipment containing modulator and demodulator, operator console with control capabilities, editing bay, and telephone for network operation
User terminals for personal mobile telephone, fax, and data two-way communications (GEO and non-GEO MSS, bent pipe or processing, L and S band)	Handheld devices similar to first- or second-generation cellular units	Transmit and receive omnidirectional antenna; all RF electronics packaged inside handheld unit as in cellphone, may include dual band/mode feature	Also contained within handheld unit, implemented in DSP and custom VLSI with high degree of integration to reduce manufacturing costs and provide reliable and user-friendly operation, facility for user authentication, roaming, reacquisition after a dropout

Table 9.1 (continued)

Application (Satellite System)	Terminal Type	RF Terminal	Baseband and User Interface
Hybrid user terminals for multimedia and Internet access (GEO and non-GEO FSS, bent pipe or processing, Ku and Ka band)	Individual fixed earth station at home or small office/home office	Transmit and receive, fixed dish (60 cm to 1.8m) for GEO, tracking dish, or phased array for non-GEO; outdoor electronics with SSPA module	Digital indoor unit containing modem, multiple access controller, and interface port cards; includes secure conditional access and monitoring for billing and network management

offices and retail stores. A user terminal intended for use by military, police, or safety services must be designed to satisfy security and durability requirements.

A final category of user terminal is the handheld variety currently available for MSS systems. Because of the unique requirements for handheld terminals, we address this subject in Section 9.6. In the following sections, we dissect the typical user terminal in order to gain a better understanding of what makes it "tick".

9.1 General Configurations

We will use the interactive VSAT as a general model of a user terminal, shown in block diagram form in Figure 9.1. This configuration allows us to discuss both the receive and transmit functions in much the same way as we did in Chapters 2 and 3. Of course, a greater degree of integration and economy is part and parcel to this kind of facility, and thus many functions are either combined or not present in the first place (or only as needed for the intended use.) Receive-only terminals used for various DTH services would not have a transmit capability (this is the most direct reason why it is difficult to add a transmit function to a standardized receive-only terminal). What we will quickly see is that transmit function is the key enhancement in the interactive VSAT or multimedia terminal (and why it has, for the time being, been so costly in comparison with TV receive-only, or TVRO, terminals). Another important aspect is the hardware and software architecture of the baseband section. With the widespread use of digital signal processing and

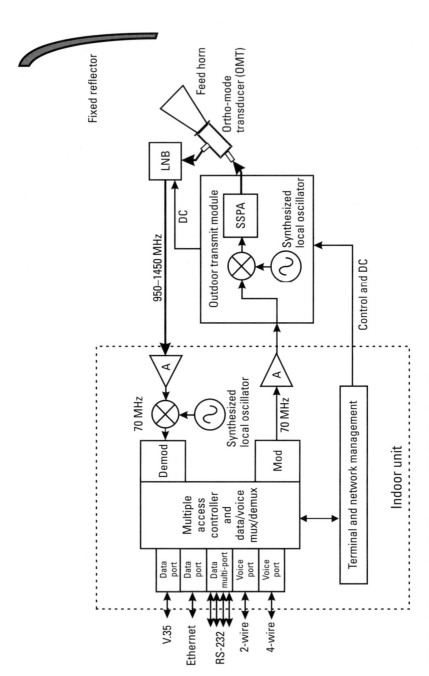

Figure 9.1 Block diagram of typical two-way VSAT for fixed satellite communications.

fast computation under control of embedded and downloadable software, user terminals are employing fewer discrete and analog components. This is introduced in Section 9.2.4 under the general topic of software-defined radio (SDR).

9.1.1   RF Terminal

Much of the terminal cost in past years has been concentrated in the RF section, which sends and receives microwave power over the air interface. These are analog components that must tolerate an external environment under all operating conditions. Within time and the introduction of low-cost designs and mass production, RF terminals are no longer the most costly element in many cases. In Figure 9.1, we see the antenna system and feed, low noise block converter (LNB), upconverter, and power amplifier. These are familiar elements in earth stations and satellites as well. The antenna and feed combine to produce the desired gain, directivity, and polarization performance (linear or circular). Because the RF terminal provides the interface between IF and RF, it could be fairly generic and possibly be used for a variety of applications. The design driver is the link budget, which in turn depends on the frequency, power, bandwidth, modulation, and multiple access characteristics of the satellite and network.

The electronic elements are usually contained in the RF head, an integrated package with the essential elements of the outdoor portion of the terminal. The primary function of the RF head is to deliver the necessary downlink, and in the case of two-way transmission, the uplink as well. Sizing of the power amplifier follows directly the procedure provided in Chapters 3 and 7. Examples of the critical elements are covered in a subsequent section. An interfacility link (IFL) cable carries signals and power between the indoor unit and the RF head. Other functions such as control of the RF head and electrical power for heating of the reflector (e.g., deicing) are provided by the IFL, when required. Installation considerations for the RF terminal and IFL depend on where the system is to be located and other aspects of the application (Section 9.5.3).

9.1.2   Baseband

The baseband section of the user terminal would normally be part of the indoor unit, as illustrated in Figure 9.1. We see the modem, multiple-access controller, multiplexing, and channelization functions. These are configured for the service needs of the overall network design and functionality. Much of the processing within the indoor unit is performed by DSP and application-specific integrated circuit (ASIC) devices (this is no different from commercial set-top boxes and cellular telephones, which are mass-market devices that perform to amazingly high levels). The service from consumer-type fixed interactive terminals would be delivered and controlled from a central hub or network control facility to monitor use and provide service management and billing. These aspects are already part of the GMPCS systems like GlobalStar and ICO, and DTH services like DIRECTV and SKY; transfer of this technology to interactive fixed services is still in the offing at the time of this writing. One of the complications will be the massive software systems that one needs to manage this type of business, something that current terrestrial operations (such as Internet service providers and cable companies providing cable modems) are only now coming to grips with. This is an important issue because subscribers cannot be acquired and signed on, and revenues cannot be collected, if the software and necessary personnel are not in place and working effectively as an integrated system.

Now that many businesses have moved to e-commerce, the ability of users to place direct service orders and changes to their service packages over the Internet becomes an imperative. This can be an advantage to all parties because there is less opportunity for misunderstanding between user and customer service representative, and the resulting action is implemented almost seamlessly. In DTH and other content-delivery systems, the conditional access (CA) arrangements combine a strong encryption service along with ability of the service provisioning system to configure individual users for their unique set of access services and applications. These involve point-to-multipoint delivery on a broadcast basis as well as greater dependence of two-way interactive features now appearing. Satellite networks have some special advantages in this regard because CA data propagates directly from the service provider to end user without passing through other networks beyond the provider's control.

9.1.3   User Interface

The interface with the end-user device or person (depending on the application) is the part of the user terminal most driven by the service. In the case of a self-contained terminal such as an MSS telephone, there may not be an interface per se. Application of VSATs and DTH set-top boxes demand appropriate user interfaces that would replace a terrestrial communication device or service. The requirement then is to represent a standard interface sufficiently that the end-user device (PC, TV set, or telephone, as appropriate) is served in as seamless a manner as possible. It goes back to the old adage in the PC world, to wit "plug and play." Done right, any end user (possibly a consumer) should be able to connect the satellite terminal interface directly into a common user device, turn the equipment on, and get to business (or play, as the case may be).

The system designer has the task of specifying the functionality to be placed within the interface as opposed to the rest of the baseband section. Like a desktop PC, the terminal might be assembled from standard modules, the user interface being one of them. This has been the tradition in VSAT design because of the significant differences among networks for particular companies or government organizations. That is, while the VSAT is constructed from standardized modules, no two networks and hence configurations would be exactly the same (this can be in terms of the software as well as the hardware). A multimedia network operating at Ka band is more of an integrated offering to a larger customer base, not unlike DTH, and so the modular approach need not be pursued. Tighter overall integration probably allows more reliable hardware and software performance. For example, software would be maintained in read-only memory (ROM), which is more reliable than if it were stored in RAM or on a local hard drive. Software updates, when needed, would be delivered in the background over the satellite network.

The previous discussion was generic, as we were speaking of the overall requirements and allocation of functions. In the following sections, we consider the details of each of the major sections, drawing on the fundamentals presented in the previous sections.

9.2 Antennas for User Terminals

Antennas for user terminals are electrically like those applied to major earth stations. The differences are mainly in terms of size (they must be physically smaller to permit simplified installation and use) and cost (which is usually very low to complement that of the rest of the terminal). The resulting lower gain and broader beamwidth must be taken into account in the system design and, specifically, the link budget. Antennas that operate at the lowest extreme of size and gain may be restricted in terms of their ability to communicate through a bent-pipe satellite with other terminals of like size. For example, VSAT antennas in the 1.2-to-1.8-m range at Ku band would normally be used in a star network with all links terminating at the much larger hub station. An exception to the rule would be if the satellite could compensate for low uplink EIRP through high-gain spot beams or a regenerative repeater, or a combination thereof. The following sections provide a review of user terminal antennas, divided into directional designs and omnidirectional antennas, e.g., those with low gain and no directionality.

9.2.1   Directional Antennas

The prime focus fed parabolic reflector antenna is the dominant design for user terminals in satellite communications. As illustrated in Figure 9.2, offset geometry is favored because of reduced sidelobe radiation from energy scatter by the feed and feed support. This design also permits rain and snow to drain

Figure 9.2 Prime focus fed offset reflector antenna as applied to a transmit/receive VSAT (photo courtesy of Newtec).

more easily, a natural result of the more acute angle of the feed. Gain and beamwidth are determined by the diameter and operating frequency, as discussed in Chapter 3. Additionally, smaller reflector size assures minimum gain loss due to pointing error and surface accuracy. An exception might be for extremely cheap designs offered at C band for TVRO applications (some have actually been made of poorly shaped chicken wire). Operation at Ka band will place greater demands on the reflector for surface accuracy. In general, stamped or spun solid aluminum reflectors provide good performance for microwave frequencies through Ka band.

Feedhorn design must take account of the required radiation distribution across the reflector, which is a function of the ratio of focal length to diameter (F/D) and the frequency of operation. While typical antennas operate with a single beam from one feedhorn, DTH operators in the United States and Europe have introduced dual feed/beam designs to allow simultaneous reception from two close-by GEO satellite orbit positions. Including the transmit function will require an ortho-mode transducer (OMT), such as that illustrated in Figure 9.3 for linear polarization. Dual-polarized operation is introduced with a switchable polarizer. The challenge comes in low-cost hardware that operates over the required frequency band with acceptable loss. This has been achieved to date for DTH receive-only antennas that sell for under $100 with feed, LNB, and mount. Producing adequate hardware with dual-mode transmit and receive service at Ka band will place demands on designers and manufacturers.

Mounts of reflector antennas at diameters under about 3m are relatively simple and inexpensive. The key aspect is the type of structure that will anchor the antenna (see Section 9.5.3). The supporting structure must be sufficient to withstand the physical weight of the antenna to permit the beam to be aligned with the satellite in the first place and maintain that pointing without adjustment. The latter point will depend on the weather conditions in the area, the most critical of which is wind loading. In benign climates like the west coast of the United States, wind is usually not a concern and so antennas can be placed on almost any structure. However, the extremely high winds associated with typhoons make antenna installations in Hong Kong and other Pacific locations quite problematic. Here, there are two factors: whether the antenna can stay on target during the storm or whether it will survive at all (without breaking or causing harm if broken free).

An alternative to the passive reflector is the phased array consisting of several broadbeam radiating elements that are combined to produce a single directional beam. The individual elements may take any of the following forms:

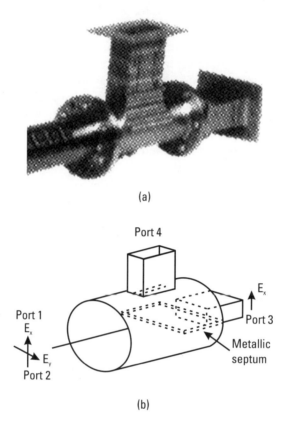

(a)

(b)

Figure 9.3 Arrangement of an OMT used to combine transmit and receive in the same feed: (a) illustration, (b) drawing.

- Half-wavelength dipole (producing linear polarization);

- Crossed dipole (which is effectively two half-wavelength dipoles phased to produce circular polarization);

- Open-ended waveguide or horn (typically with small dimensions to produce a broad beam);

- Waveguide slot (an attractive approach, since it allows the waveguide to distribute the power between slots).

An array of these elements can be formed using phase and power combining. While the efficiency of a properly designed and manufactured array antenna can be as good or better than a passive reflector, the fact that it costs more to manufacture has limited its position in the marketplace.

The electrically steered phased array promises to overcome these deficiencies through active power distribution and phasing, under software control. Each element of the array is connected to an amplifier and phase/gain distribution network to allow pointing of the beam in different directions without physically moving the antenna. This would allow the antenna to be mounted to a surface even if that surface is not perpendicular to the direction of radiation. The next step is to add the transmit function to provide an interactive type of terminal. This involves the challenge of separate uplink and downlink bands, since the two directions operate on different frequencies. Two alternative approaches could apply: install separate arrays, each optimized for the particular band (this is the common approach in spacecraft like ICO and Globalstar); or provide a diplexer for each element to separate the directions based on either polarization or frequency or a combination thereof. The advantage of the former is that the two directions never exist in the same physical structure, which greatly simplifies the electrical design. Bandpass filtering prevents energy from entering the receiver, but this can be done on a broadband basis. The disadvantage is the physical space and mounting area needed for separate arrays. If this can be acceptable, then having individual transmit and receive arrays is highly preferred. Combining transmit and receive requires a separate diplexer for each individual element and power must be coupled efficiently to it. Performance would probably be somewhat less than what one would expect to obtain with separate arrays.

9.2.2 Omnidirectional Antennas

In cases where orientation of the terminal-to-satellite link may be unpredictable, it is a general practice to use an omnidirectional antenna. Omnis are associated with non-GEO satellites that may appear anywhere in the sky. The advantage of not favoring any direction is balanced by the following characteristics:

- Omnis have low gain, approximating that of an isotropic radiator (0 dBi), but potentially negative in some directions (and positive in others, since the integrated pattern over a sphere produces the same total power).

- They lack directional properties and therefore cannot control co-channel interference from other visible satellites.

- Polarization performance may be poor as well, whether linear or circular.

A useful property of omnidirectional antennas is that they tend to be physically small, making them ideal for mobile and handheld applications.

Examples of common omnidirectional antennas are presented in Figure 9.4. The notched printed patch antennas in (a) are applied in situations where a low profile is desired. In (b), the patch is driven by a phase shifter to produce circular polarization, which is required for many of the mobile satellite systems. The crossed-dipole antenna in (c) is easily driven by the transmission line and produces a well-formed circularly polarized

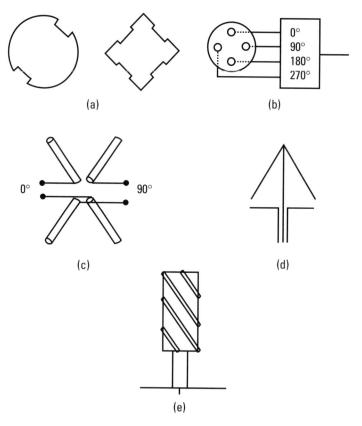

Figure 9.4 Circularly polarized omnidirectional antennas suitable for mobile satellite applications: (a) symmetric patch with perturbation notch, (b) symmetric patch with balanced feed, (c) orthogonal dipoles, (d) inverted vee (drooping dipoles), (e) quadrifilar helical antenna.

pattern. An enhanced version called the drooping crossed dipole is shown in (d) and in the photograph in Figure 9.5. The individual elements of the drooping dipole are made thick to increase bandwidth (in percentage terms). Below the drooping dipole elements can be found a metal ground plane which improves performance by providing a uniform reflector. This can be made part of the antenna, or devised from the skin of the structure that it is attached to (e.g., if it were mounted on a vehicle or aircraft). While this design is excellent in its performance it nevertheless is not common to user terminals because of its bulkiness. The radiation pattern, however, provides a good reference as to what one can obtain with a practical antenna.

Formed by winding four wires along a cylinder, the quadrifilar helix antenna ((e) in Figures 9.4 and 9.6) is very popular for handheld applications requiring omnidirectional radiation. Each of the diagonal lines represents a loosely wound turn on the same cylinder. The number of turns, length, and diameter of the assembly determine the bandwidth, radiation pattern, and impedance. The shape of the pattern above the horizontal is very similar to that of the drooping crossed dipole, with more pronounced backlobe radiation due to its lack of a reflecting groundplane. The quadrifilar helix can be designed for a variety of antenna pattern shapes, gains, and impedances.

9.3 Power Amplifiers

User terminals applied to two-way satellite communication services must be capable of producing the required uplink carrier power level, as dictated by the link budget. There is the obvious tradeoff to produce the required EIRP in terms of antenna gain versus transmitted power at the antenna port. The latter is the responsibility of the power amplifier (PA), a device that introduces the desired amount of gain, per the transmit gain budget, and conversion from either DC or AC prime power into the RF transmit power. In most cases, the PA is mounted directly to the antenna feed to reduce losses and thereby minimize the transmit power required. There are additional requirements on the amplification function (as for any amplifier), such as bandwidth, linearity, efficiency, and operating stability over the expected temperature range. Noise contribution is not a primary consideration with placement of the PA after all other gain stages.

In the following paragraphs, we review common amplifier types that can be found in a variety of user terminals, for fixed, portable, and mobile applications. We benefit by recent developments in wireless communications where new transistor technologies, circuit designs, and monolithic

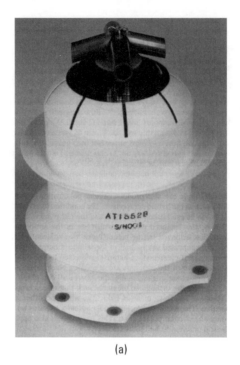

(a)

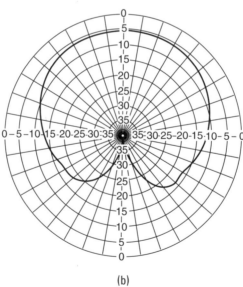

(b)

Figure 9.5 Illustration (a) and performance (b) of a drooping-dipole antenna with crossed elements to achieve circular polarization and symmetrical coverage of the sky.

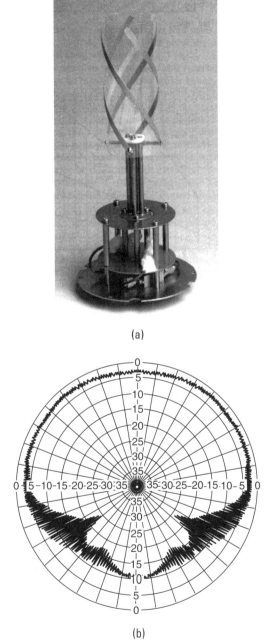

(a)

(b)

Figure 9.6 One-half turn quadrifilar helical antenna (a), indicating its symmetrical radiation pattern at 1.537.5 MHz (b).

microwave integrated circuit (MMIC—pronounced "mimic") modules have been introduced into the satellite communication marketplace. This is because many satellites operate in microwave bands that are either adjacent to or overlap the bands common to PCS/PCN, wireless local area networks (wireless LANs), and local microwave distribution service (LMDS).

Key amplifier parameters are shown in Table 9.2. Solid-state power amplifier (SSPA) designs, which simplify installation and operation, are available up to 30 GHz, with most of the current development going on in the L band (1.5 to 1.9 GHz) and Ku band (14.0 to 15.5 GHz). This is not to say that C- and Ka-band amplifiers cannot be obtained, only that it will be relatively easier to obtain competitively produced amplifiers for typical applications at L and Ku bands. For example, the introduction of Ka-band satellites is sufficient motivation for manufacturers to invest in the development of the hardware and production processes.

As with satellite power amplifiers, SSPAs have not entirely replaced TWTAs. Rather, there are many applications in the user terminal area for which only a TWTA will be appropriate. The key driver is the power output: the lower end of the power range would be approximately 200, 100, 50, and 10 watts for amplifiers in the C, X, Ku, and Ka bands. In almost no case will a TWTA be used in S and L band because of availability of SSPAs with appropriate capability. TWT amplifiers are available from CPI and NEC in the lower powers, making it possible to demonstrate Ka-band applications with the satellites already in orbit.

A useful measure of performance of wideband amplifiers is the gain-bandwidth product [1]. Assuming that the amplifier has N individual amplifier stages, then the gain-bandwidth product is:

$$(GBW)_N = (Gain)^{1/N} BW$$

where N is the number of stages, $Gain$ is the total midband gain (as a ratio), and BW is the overall bandwidth. Because amplifiers with fewer stages have better gain-bandwidth performance (and cost less as well), most applications try to work within the constraints of one or two stages in tandem. An individual stage would have the typical equivalent circuit shown in Figure 9.7. The principal bandwidth-limiting element is the input capacitance, C_{in}. The electronic operation of PAs is usually defined according to the letter designations given in Table 9.3. An example of an actual transistor circuit can be seen in Figure 9.8, which is for a Class A amplifier with stabilizing feedback

Table 9.2
Design and Performance Parameters for RF Solid-State Power Amplifiers Used in Satellite
Communication User Terminals

Parameter	Properties	Impact on Terminal	Impact on System
Power output	Between 1 and 10* watts	In combination with antenna gain, produces EIRP	Lower power preferred to minimize interference to adjacent satellites and other services
Gain	5 to 20 dB	Determined by individual stage design and number of stages	No direct impact
Bandwidth	Sufficient for carrier transmission; 50 kHz to 10 MHz	Narrowband amplifier may limit utility over time; wide-band design best for flexibility over satellite bandwidth	Determines how the terminal will operate in the network, including degree of flexibility for operation on other frequencies
Efficiency	40 to 70%, depending on design and operation	Low efficiency pushes DC or AC power requirement up, which may be acceptable for fixed installations. Mobile terminals require higher efficiency to conserve power	No direct impact
Linearity	Requirements for signal fidelity and inter-modulation will drive design	Can select from classic amplifier design alternatives: Class A amplifiers are most linear, Class B highly non-linear unless configured for push-pull. Class AB represents a compromise. Linearization also an option	Potential for adjacent channel interference or out-of-band emissions
Thermal requirements	Dissipation determined by efficiency and power level	High-frequency operation makes temperature control more difficult due to small dimensions and compactness	No direct impact

*Designs are available at C band that use a large quantity of parallel PA states to reach 200 watts of RF output or more.

between the collector output and the base input. Efficiency has two measures: the conventional DC to RF efficiency,

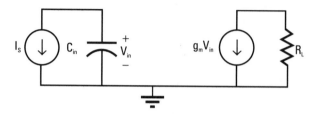

Figure 9.7 Simplified high-frequency small-signal transistor amplifier model.

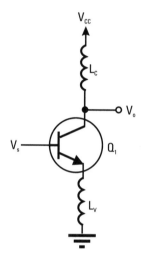

Figure 9.8 Basic Class A RF amplifier using a junction transistor.

$$\eta = \frac{P_{RF}}{P_{DC}}$$

and the power added efficiency,

$$PAE = (1 - \frac{1}{G})\frac{P_{RF}}{P_{DC}}$$

where G is the RF gain of the amplifier at the value of RF power out being achieved.

There will be situations where the gain of a single transistor stage is insufficient to meet the gain budget requirement of the user terminal. A direct approach is to employ multiple stages; however, as these equations

Table 9.3
Characteristics of Classes of RF Power Amplifiers

Class	Waveform	DC-to-RF Efficiency	Linearity	Comments
A	Full cycle (positive and negative going)	Low: 25 to 50%	Linearity depending on bias point; optimum possible	Lowest efficiency and best linearity; consumes power even if no signal present at input
B	Half cycle (positive going only)	Medium: 30 to 55%	Nonlinear due to negative clipping and operation across zero conduction	Good efficiency at the sacrifice of linearity. Does not conduct if no signal present. Push-pull operation of balanced pair of Class B amplifiers alleviates most of the problems of this type
C	Partial cycle (positive going only)	High: 50 to 80%	Very nonlinear. Only acceptable for constant envelope signals	Not popular for most satellite applications
AB	Full positive half cycle with partial negative cycle	Medium: 35 to 60%	Moderate linearity. Conducts without signal present	Push-pull operation yields good linearity with good efficiency.

indicate, the more stages we employ, the less bandwidth we might expect to deliver. This is not a problem for elevated frequencies like Ku band, where the required 500 MHz of operating bandwidth is a relatively small percentage of the center frequency. Operation at L and S band might be less satisfactory, but this can be addressed in the overall design of the terminal. For example, employing a higher power mixer or other input stage can reduce the gain requirement from the final PA.

Low-power consumption PAs of the type found in portable and handheld terminals are limited in their RF output to about 10 watts at Ku band and 1 watt at Ka band. Efficiency of DC to therefore conversion is a major concern. Another important consideration is linearity, since this impacts the end-to-end quality of the link and the possibility of adjacent channel and out-of-band interference. There is a tradeoff among these parameters, as evident from Table 9.3: better linearity means lower efficiency or power output. Constant envelope signals like minimum shift keying can tolerate a certain degree of nonlinearity, but signals with significant peak to average variation like standard QPSK with pulse shaping require a somewhat linear transmitter.

As shown in Table 9.3, the most linear type of design is achieved with the Class A amplifier. By biasing the transfer characteristic correctly, the Class A can produce near optimum performance. Enhancement is possible using a predistortion technique, also referred to as linearization. The drawback of Class A is the lower efficiency that one expects from a transistor circuit that is always conducting with or without signal present. An example of a stable Class A transistor amplifier is shown in Figure 9.8 (a junction type transistor is shown for illustrative purposes; alternatively an FET could be substituted). If one needs high efficiency, then Class B and AB would normally be the standard for design. Class B, in particular, has the added feature of not drawing current when no signal is present. An example of a Class B push-pull amplifier is provided in Figure 9.9.

Transistors for microwave amplifiers are not purchased as discrete components but rather are assembled into modules for ease of design, manufacture, and operation. Importantly, the package provides reliable connection points into the active sections of the amplifier, appropriate conduction to spread heat across the base, and protection from vibration and moisture. Examples of typical microwave power modules, using MMIC and GaAsFET technology, are shown in Figure 9.10. The following are top-level characteristics of these packaged amplifiers.

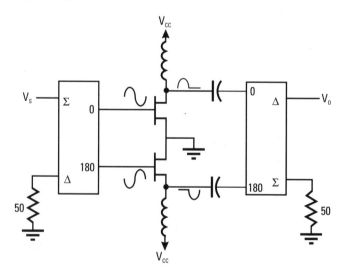

Figure 9.9 Push-pull Class B RF amplifier using a pair of microwave FET amplifier modules and two 180° hybrids.

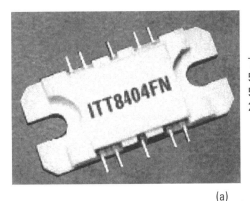

Two-stage GaAs MMIC
5.5 to 7.0 GHz
5 W output, 18 dB linear gain
25% power added efficiency

(a)

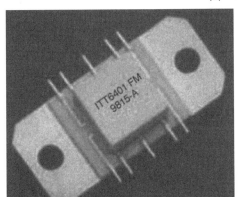

Two-stage GaAs MMIC
4.5 to 7.1 GHz
2.5 W output, 18 dB linear gain
40% power added efficiency
43 dBm third order intercept

(b)

Three-stage MESFET
1800 to 2000 MHz (PCS, DECT, and WLL)
0.5 W output, 29.3 dB gain
49% power added efficiency

(c)

Figure 9.10 Examples of typical microwave SSPA modules: (a) 6 GHz 5 watt, (b) 5–7 GHz 2.5 watt, (c) 2 GHz 0.5 watt (images courtesy of ITT).

- ITT model 8404FN is a medium-bandwidth C-band power module with two stages, producing 18 dB of linear gain in the 5.5 to 7.0 GHz range. The output power is nominally 5W into a 50 ohm load and a gain flatness of ±1.5 dB. The first harmonic is 20 dB down from the carrier.

- ITT model 6401FM, a broadband two-stage C-band power module covering 4.5 to 7.1 GHz. Power output is 2.5W at a gain of 16 dB (18 dB linear gain), for a power added efficiency of 40% (Class B, nominal). The output is matched to 50 ohms and is intended for fixed installations such as VSAT earth stations.

- ITT model 2205AF, a 0.5W RF output L-band amplifier comprised of three stages. Operating range: 1,800 to 2,000 MHz, with typical efficiency (PAE) of 49% in Class B. The gain is 29.3 dB at 1,900 MHz and 3.6V supply voltage. The second harmonic of the carrier is 34 dB below the carrier. The unit will operate in a high VSWR environment of 8:1, indicative of severe RF reflections back from the antenna.

Figure 9.11 provides an example of a high-efficiency Ka module with approximately one-watt output developed under the sponsorship of NASA Lewis Research Center [2]. The integrated circuit (IC) was fabricated at TRW's GaAs IC foundry in Manhattan Beach, California. It uses a high-voltage source at 24 volts to achieve a power-added efficiency of 32% at 0.81 watts into 50 ohms. The mounted chip and transfer characteristics are shown in Figures 9.12 and 9.13, respectively. The gain of the amplifier is approximately 19 dB in a bandwidth of 2 GHz.

The RF head of an interactive terminal must include a low noise block converter (LNB) and an upconverter. This type of design is optimum for VSATs and other permanent installations (including onboard ships, aircraft, and vehicles). An important objective of LNB design is to achieve the lowest system noise temperature while providing the required gain for the downlink gain budget. Laptop and handheld terminals may in fact use a lower intermediate frequency to be consistent with the operation of the demodulator contained within the same piece of equipment. On the transmit side, the upconverter is attached to or even part of the power amplifier.

The PA section of the user terminal can be constructed from power amplifier modules of the type just described. A suitable power supply will often need to be included in the package. It is also possible to integrate the PA with the modulator and upconverter, producing a complete transmit

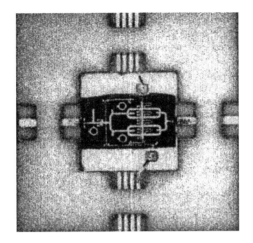

Two stages with output stage containing
four GaAs transistors in series bias,
31 to 33 GHz (Ka band uplink)
800 mW maximum output at 24 VDC
32 to 35% power added efficiency

Figure 9.11 High-voltage Ka-band MMIC mounted in a metal ceramic package. (Reference [2], with permission.)

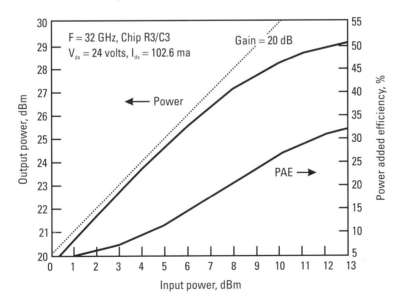

Figure 9.12 Power-frequency response of the packaged MMIC chip. (Reference [2], with permission.)

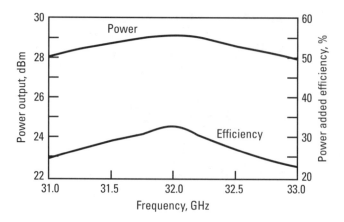

Figure 9.13 Output power and efficiency of the packaged MMIC chip at 32 GHz. (Reference [2], with permission.)

section under one case. The interface back to the user side would be the baseband signals. The following section reviews digital signal processing technologies that have been applied to the baseband section, a technique that has recently been described as software-defined radio.

9.4 Baseband Functions

While the RF elements are critical to the link budget, the baseband functions of a user terminal provide the essential information processing and access control that are central to an effective end-user service. In the past, user terminals relied heavily on analog technology, particularly baseband shaping (e.g., preemphasis and companding), frequency division multiplexing, FM reception, and possibly transmission as well. Implementation of this functionality was possible with analog ICs as well as discrete transistors and other electronic components. The circuit boards could be assembled manually or by machine to reduce costs. One difficulty of the analog approach was the variation in performance and quality from board to board, requiring custom tuning at the factory and retuning from time to time when in operation (a source of network service variation and possible instability). Terminals of this type are no longer in production.

The baseband section of the terminal insures that the user input and output signals are properly formatted for the particular application and satellite network. Involved are such functions as analog-to-digital (A/D) conversion, compression, encryption, and multiplexing of data streams for information transfer and network control. As discussed for the DTH receiver, these functions have become highly integrated with the hardware and software of the user terminal, and may be contained within a single IC. As suggested, the baseband section would control access by the end user and application, corresponding to the particular type of device (e.g., telephone, PC, TV receiver, etc.) and the manner of operation of the overall service. This could be transparent (as in the case of telephone service or Internet access) or tightly coordinated so that every element from end to end provides a necessary function (as in the case of DTH and MSS services).

The circuits to perform analog functions were less complex than today's digital processing and software techniques that one finds in all varieties of user terminals. With transistor counts in the millions and software requiring MBytes of ROM and RAM, we see wide adoption of very large scale integration (VLSI) and application-specific integrated circuits (ASICs). This allows the developer to create highly effective baseband systems at relatively low cost, provided that sufficiently large quantities are manufactured and sold. Another important innovation is the digital signal processor (DSP), a class of high-speed chip that can rapidly perform complex mathematical operations akin to the math coprocessor found in many PCs.

The baseband section of today's complex user terminal is therefore a customized high-speed computer with the software needed to create the signal processing, multiple access structure, and user interface required by the system design. To aid in our review of the capability and approach, Figure 9.14 provides the block diagram of a generic user terminal of the type that might be used for interactive two-way communications for voice and data service using CDMA [3]. The primary objectives of this design are low power consumption and minimum transistor count. Shown on the right is the RF terminal section consisting of the antenna, HPA and LNA, up- and downconverters, and frequency synthesizer (used to select the desired channel frequency). Analog-to-digital and digital-to-analog converters (A/D and D/A, respectively) are indicated as discrete elements that could either be external to or part of the digital processing section. The latter is shown within the box, and consists of CDMA spread and despread functions, as well as the modem. Timing and digital control are also part of this section of the terminal, necessary for synchronization at several levels.

Input and output on the right side of the figure is by way of a voice or data code, which provides compression, service control, and encryption (as required). Compression may be treated as a separate module that might be

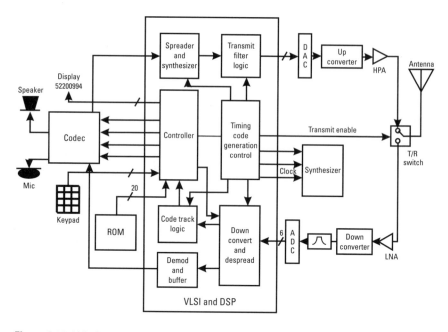

Figure 9.14 Wireless digital user terminal employing a spread-spectrum receiver.

obtained from a specialist source. The antenna on the right happens to be connected to the RF electronics through a transmit/receive switch, allowing only one or the other function to occur at the same time. Most satellite user terminals allow simultaneous transmission and reception through a common antenna using a diplexer.

9.4.1 Software Design Process

The first step in the design process is to create a computer model of the baseband functions using an appropriate software design tool. An attractive approach is to use a signal analysis tool like SystemView, which allows the user to first analyze the performance of the theoretical configuration prior to creating the actual software design. What this means is that the engineer can build a visual model in block diagram form on the computer screen, simulate the signal flow through the model, examine the results, and draw inferences about next steps. This is an iterative process that should produce a workable model of the actual circuit. Then, the tool will output the definition of the very large scale integration (VLSI) circuits and software code to actually implement the baseband functions in silicon. This definition is typically expressed in a programming language called VLSI Hardware Definition Language (VHDL). It is a formal notation intended for use in all phases of the creation of electronic systems. Because it is both machine readable and human readable, it supports the development, verification, synthesis, and testing of hardware designs.

An example of a popular methodology is that offered by Synopsys of Mountain View, California, one of the leaders in the area of DSP design and verification tools. The following is a typical flow of the process:

1. Specify your design (block diagram, signal flow, allowable impairments, operating constraints, etc.).

2. Set performance, area, environment constraints, and optionally, power.

3. Perform translation and architectural optimization of VHDL components.

4. Compile the VHDL code.

5. Select a technology library supplied by the desired semiconductor vendor.

6. Optimize the design under various speeds, area, and power constraints and select the implementation that best meets needs.

In many cases, the approach of using a single tool does not quite satisfy all of the requirements, including the flexibility to handle many modes of transmission and different codes or multiple access methods (e.g., for a dual-mode terminal). The designer will therefore have to resort to working one level down in the hardware/software design, using more powerful (but less friendly) software tools. These are available from Synopsys and other specialists.

9.4.2 Software-Defined Radio Applications

Theoretical advances in communications systems predate their implementation in the commercial world because technology was limited to what could be created with hardware elements. With the adoption of DSP, complex software tools and techniques, and VSLI, there are much greater possibilities for using all of the advances of the last decades plus new ones as they appear. The greatest power of these technologies might have more to do with rapid application development and quick time to market than with the fact that more circuits can be put on one chip (although this is certainly important as well).

Advanced by Brad Brannon of Analog Devices, software-defined radio (SDR) is a new paradigm in hardware and software and as a consequence has many potential uses in satellite and terrestrial wireless communications. Through software control or download, an SDR can change its frequency, bandwidth, and modulation scheme. The alternate of a hardware reconfiguration would only provide a limited range of flexibility. True software programmability offers a range of operations that individually can be highly tailored to a specific application. To offer a range of operation, several things must happen. First, data must be manipulated with a high-speed digital signal processor (DSP), which forms the heart of a software radio. In addition to a general-purpose DSP, numeric accelerators are required to process the carriers digitally.

The following discussion uses a particular terminal design as a basis to review a software-based implementation approach. The receiver front end consists of an analog RF downconverter that converts the desired signal band to a convenient IF frequency for digitization (Figure 9.15). The downconverter is followed by a high performance A/D converter, which digitizes the entire IF spectrum bandwidth (typically 30 MHz). The A/D converter output is processed with a receive signal processor (RSP) which is responsible for individual channel selection and filtering. The output of the RSP consists of a filtered digital IF signal requiring only demodulation by the DSP. The RSP is a numeric preprocessor that can replace a local oscillator (LO), quadrature mixer, channel select filter, and data decimation filter (Figure 9.16). Decimation is the process of reducing the quantity of data per unit time by passing a fixed percentage of samples and neglecting the rest (e.g., using 10% and discarding 90%). In a typical application, one RSP provides the necessary channel selection and tuning functions. The channel characteristics that are programmable may include data rates, channel bandwidth, and channel shape.

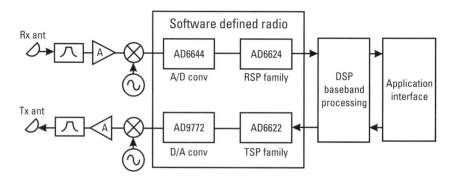

Figure 9.15 Software-defined radios combine the functionality of A/D conversion, receive signal processing (RSP), transmit signal processing (TSP), and D/A conversion.

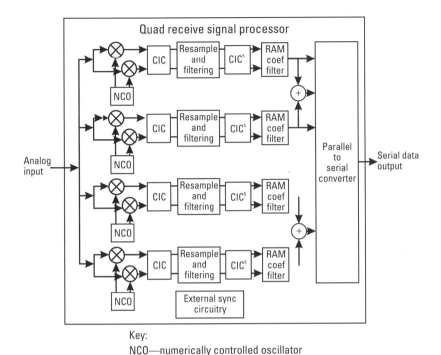

Key:
NCO—numerically controlled oscillator
CIC—Cascaded integrated comb filter

Figure 9.16 In a software-defined radio, the receive signal processor (RSP) is designed to replace an LO, quadrature mixer, channel filter, and data decimator.

To function correctly, the SDR device must be capable of handling the data rates required by the A/D converter interface. Considering the number of concurrent operations, the RSP may be processing data at a sustained rate of several billion operations per second (BOPS), which includes the numerically controlled oscillator (NCO), quadrature mixer, decimation filters, and fast-input response (FIR) filters. The next specification of interest is the internal parallel data bus width, which must be wide enough (contain enough bits) to preserve signal integrity. Although the A/D converter may only be 14 bits wide, oversampling followed by narrowband filtering improves the effective signal-to-noise ratio (SNR); this processing gain can amount to up to 30 dB for the channelization scheme of the air interface. This is the equivalent of 5 more bits. Therefore, internal bus widths must have the equivalent of at least 19 bits to preserve signal integrity.

It is important that the RSP include flexible decimation and filtering configurations that allow for a wide variety of data rates and filter bandwidths. One of the distinct features of an RSP is the ability to select the desired analog frequency very precisely. Most RSPs have a 32-bit NCO, which provides a frequency resolution of about 1 in 4 billion. This is usually much more than adequate, but gives great flexibility. In addition to the flexibility, frequency hopping is greatly simplified. Since a phase-locked loop (PLL) is not used, changing frequencies is instantaneous. This can be a great benefit in time-division applications, such as TDMA, where hopping must occur within the guardband of a few Hz. The transmit signal processor (TSP) is a numeric postprocessor that replaces the first LO, quadrature modulator, channel filtering, and data interpolation. As with the RSP, the TSP sets the transmitter apart from traditional designs because all channel characteristics are now programmable (e.g., data rate, channel bandwidth, and channel shape).

There are several specifications that are important when selecting a TSP. First, the device must be capable of generating data at the rates to preserve Nyquist sampling over the spectrum of interest. Therefore, the TSP must be capable of generating data at least twice as fast as the band of interest and preferably three times faster as reasoned earlier for anti-aliasing filter response. Similar to the RSP, the TSP's bus widths are important, yet for different reasons. If the TSP is used in a single-channel mode, then the issue is simply quantization noise. It is usually not desirable to transmit excess in-band (or out-of-band) noise as this wastes valuable transmitter efficiency.

In a multicarrier application, many channels would be digitally summed before reconstruction with a D/A converter. Each time the number of channels is doubled, an additional bit should be added so that dynamic range is not taken from one channel when another is added. As a general rule, the bit precision of the TSP + D/A converter should not be a major contributor to the noise figure of the transmitter. Again, channel bandwidth and shape are important and require flexible interpolation and filtering options. Since a TSP implements frequency control with an NCO and mixer, frequency hopping can be very quick, allowing implementation of the most demanding hopping applications.

Timing design and testing is an area of critical importance in the design of SDR systems. The tools available to designers allow them to create a functional structure and then identify timing problems quickly [4]. This must be done at the chip level, followed by a similar exercise at the integrated unit level. The objective is to develop a complete product that performs like it was formed from discrete components. Once the circuit and system are defined, the next step is to verify that timing relationships are correct and that the combination performs as near to expected as possible. This usually involves much iteration with the hardware and software, using a variety of tools and software testing methods.

One final component necessary to develop a proper software-defined radio is the D/A converter. The first specification of interest is signal-to-noise ratio (SNR). As with any A/D converter, SNR is primarily determined by quantization and thermal noise. If either is too large, then the noise figure of the D/A converter will dominate the overall signal chain noise. Another important specification for the D/A converter is spurious response. In the worst-case scenario, out-of-band spurious emissions may violate local or international rules and regulations. This impacts the choice of sampling rate and number of bits per sample. Most air interface standards require approximately 65 dB rejection of spurs at the antenna. In allowing for margin and degradation in the following stages, typical D/A converter requirements dictate that such spurs be 75 dB below the carrier (a difficult specification according to traditional analog design principles).

This discussion was meant to introduce readers to the approach and methodologies in practice for creating hardware and software baseband machines. Representing the result of many years of investigation and practice in digital signal processing and complex IC development, one cannot expect to put the complete story down for all time. Instead, we hope to instill a useful notion of the style and capability of the design process.

9.5 Fixed Terminals

Fixed user terminals are intended for permanent or semipermanent installations on buildings and other stationary locations. They typically include some type of directional antenna system which is itself fixed in orientation; alternatively, the antenna may have a tracking system to maintain pointing on satellites (a requirement in non-GEO operations). The electronics within the terminal normally would use prime AC power of the type locally available. This can be a concern, since power supplies vary from country to country and even region to region. For this reason, modern user terminals (like laptop PCs) are designed to operate from any AC voltage in the range of 100 to 240 VAC. A battery and charger may be included to provide a complete uninterruptible power system (UPS) that isolates the circuitry from power variations. These features make user terminal designs more practical and will enhance their adoption around the world. As we have discussed above, the electronics are configured into the RF terminal that is associated with the antenna and the indoor electronics that provide the modem and baseband functions. Examples of typical fixed terminal classes and their particular requirements are discussed below.

9.5.1   Receive-Only (TVRO)

The receive-only class of user terminal has been with us since Neiman Marcus offered the first home dish system back in 1977. Costing nearly $20,000, this home TV system was definitely an oddity; today, literally any TV home can afford the type of TV receive-only (TVRO) terminal that has proliferated throughout the Americas, Europe, and Asia. While the first units were capable of receiving one FM TV channel within a transponder, the digital models allow the user to recover 200 or more digitized channels (using the MPEG format) from multiple satellites operating in a single orbit position. Ancillary services like pay per view (PPV), onscreen program guide, music, and Internet-based data broadcasting have become available on a wide scale. The technology that facilitates these services is the artful combination of traditional RF terminal components with digital processing commonly found in mobile phones and the home PC. The difference here is that the downstream bandwidth and data rates are measurable in the Gbps instead of the kilobits- or megabits-per-second range. Broadband satellite communications are already commonplace, provided on a broadcast if not interactive basis (interactive features have already been demonstrated at the time of this writing and will be discussed in the next section).

A basic block diagram of a digital TVRO is provided in Figure 9.17, based on the configuration for DIRECTV Satellite System (DSS). Capable of receiving any of the 32 transponder channels within the 12.2-to-12.7-GHz range of the Region 2 BSS plan, this receiver can demodulate the selected carrier to recover one of N (e.g., up to 12) digitally encoded MPEG streams. The simple RF terminal is composed of a 45-cm solid reflector on an azimuth over elevation mount, a circularly polarized dual feed and low noise block converter (LNB). Channel capacity was increased further with a dual feed design to allow reception from a second orbit position seperated by 10 to 20° (the range is limited by the gain loss from scanning away from the reflector centerline). The output of the LNB is the full 500-to-1,200-MHz bandwidth in either the right-hand circular polarized (RHP) or left hand circular polarized (LHP) downlink, delivered in an IF range of 950 to 2,150 MHz. This allows the use of low-cost coaxial cable to carry signals to the set-top box inside the home. Selection of RHP or LHP is achieved electrically within the feed polarizer; alternatively, the feed can be fitted with two LNBs to permit both polarizations to be carried

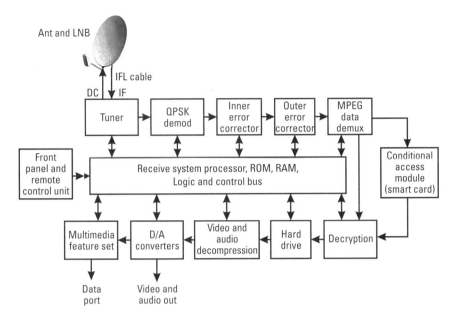

Figure 9.17 Generic block diagram of a digital direct-to-home integrated receiver-decoder (MPEG-2).

indoors. In this way, there can be multiple set-top boxes within the same dwelling. An additional LNB set is required for a second feed, if present. DCs voltage over the cable powers the LNBs and controls selection of received polarization.

The set-top box provides all of the facilities to select channels, control operation, and deliver analog video and audio to the TV receiver in either analog or digital (MPEG-2) formats. The analog tuner section of the box selects the appropriate satellite transponder channel according to the channel desired for viewing. Demodulation of this carrier produces a digital baseband containing errors from the uplink and downlink. Most of these errors are corrected through a combination of the outer and inner forward correction codes, allowing an essentially pure stream of data to reach the demux section of the receiver. The purpose of the demux is to select the desired TV channel from the combined data stream and provide the data to control operation and service to other end-user applications (if provided).

Decryption is performed under control of the conditional access module using the key provided by a subscriber interface module smart card. Control is further exercised over the satellite link with data that was inserted by the subscriber management center of the DTH service provider. In this manner, the selection of programs is delivered according to the particular subscriber's service package and service-level agreement.

The output of decryption is the multiplex stream of MPEG-formatted video and audio channels, along with data in various forms. The latter includes the control information needed to activate the subscriber's service mix, provide the onscreen program guide, and control the overall operation of the box. In particular, the unit includes a dedicated microprocessor (shown at the center of Figure 9.17) that also includes ROM and RAM memory. The last step in the video chain is the D/A converter that delivers a conventional analog TV signal modulated to a standard VHF channel (e.g., 3 or 4 in North America). Alternatively, the signal can be made available as a composite signal at baseband. Additional features could include an internal hard disk drive to allow pausing and recording for later viewing, and a multimedia or PC port through an Ethernet connection. DIRECTV requires a telephone connection as well if the subscriber wishes access to pay-per-view services (e.g., near video-on-demand).

The principal user interface is a remote control unit (or channel changer). The ones supplied by Thomson Consumer Electronics under the RCA brand name were universal remote controllers that could be programmed to control the TV set and VCR, in addition to the set-top box. The front panel of the box is usually very clean, including a power-on light and a few buttons for a limited set of functions. The back of the unit provides a variety of sockets and connectors, primarily for the connections described previously. One of the more controversial has been for an external TV antenna to receive local over-the-air signals (alternatively, the connection can be used for cable TV). This would seem to be counter to the intent of using a DTH service in the first place, but has been necessitated by the lack of local channels available over the satellite system. Recently, the major players in the United States have introduced local channels to the top 25 markets. User terminals of this type have been manufactured in the millions and represent the first true consumer product within the satellite communication industry. They can be purchased for as little as $150 (including antenna and LNB) and be installed in a few hours (less if the cable is already available to connect between antenna and set-top box).

9.5.2 Fixed Transmit-Receive Terminal (VSAT)

The standard term for a transmit-receive earth station with a relatively small fixed dish is the very small aperture terminal (VSAT). VSAT systems with antennas as small as 1m are installed at filling stations, office supply stores, and branch offices of brokerage firms. VSATs are largely intended for enterprise and other private data communication applications, and their use for public network services has been limited. As such, they are "industrial" systems with price tags to match (e.g., about $4,000 per installation). This is beyond the means of the average consumer, but products have appeared which begin to match what has been done in the DTH arena. This aspect is considered at the end of this section.

A block diagram of a typical data and voice VSAT terminal is shown in Figure 9.1. The RF terminal section, shown at the right, is composed of an LNB, upconverter, and SSPA. Reflectors in the range of 1m to 2.4m are common, depending on the satellite design and frequency of use. For example, with a Ku-band satellite capable of about 50 dBW of saturated EIRP and a G/T of greater than 3 dB/K, the lower end of the diameter range is feasible. However, if C-band operation is required, then the lower EIRP footprint level (generally less than 40 dBW) and close spacing of the satellites will push the needed diameter up to 2.4m in the typical installation. Sizes in the middle, as determined by the link budget, will allow for the particular data rate, satellite performance, frequency band, and, in the case of Ku or Ka band, the particular rainfall region where the terminal is located.

In almost all cases for standard VSATs, the indoor electronics are housed in a separate chassis about the size of a PC. This allows the unit to be placed within a computer room or even a typical office that has access to the computer and telephone systems to be supported. Port cards, shown to the left side of the indoor unit, provide standard interfaces for the locally attached devices. For example, one or more of the data ports could accommodate an Ethernet attachment at 10 Mbps or 100 Mbps; another type of data port supports a direct connection via an RS-232 cable to a PC for communication with an Internet service provider (ISP). Voice ports are used along with a telephone system such as a private branch exchange (PBX) or key system (the latter is basically a scaled-down PBX). Another approach is to use the VSAT indoor unit as a wireless telephone.

The electronics and software within the indoor unit can be configured for the desired services. However, the air interface is driven by the particular data rates, modulation, and multiples access that are employed over the link. We see in Figure 9.1 the channel control elements of the terminal that support the multiple access system (FDMA, TDMA or CDMA), the modulator and demodulator, and the interfacility link (IFL) cable between the indoor unit and the RF electronics. This cable also provides DC power as well as a means to monitor and control the outdoor electronics.

The consumer version of the VSAT can be exemplified by the DTH home receiver system, composed of a low-cost outdoor antenna and RF package, a cabling system into the home office or other room, and either a compact STB or printed circuit board to be installed in a PC. This is similar to the terminal adapter (TA) card that is required for integrated services digital networks (ISDNs) provided through the digital public telephone network. There would also be special software to operate the board and perform supporting functions associated with preparing the data for transmission over the satellite link. The objective is to make the operation as nearly transparent as possible to the user.

9.5.3   VSAT Installation Considerations

User terminals for fixed applications must, of necessity, be easy to install and operate, providing an unobstructed line of sight to the satellite or satellites that provide the service. This is the most important installation requirement, but it is not the only one. As suggested below, the task may be challenging in terms of time and cost. This is why it is normal practice to perform a site survey before starting the installation project, taking into account the following:

- *Orbit visibility*—ability for line of sight toward the desired orbit position; consideration should be given to pointing at other positions that might be usable in case of a satellite change; for non-GEO satellite constellations, the entire sky should be viewable above a minimum elevation (e.g., 20°).

- *Frequency clearance* (shared band)—suppression of potential RF interference with terrestrial microwave stations operating on the same frequency (a concern at C band).

- *Available real estate*—adequate physical space for the antenna, its mount, and any supporting equipment or cables. Along with the next item, this requirement often leads to negotiation with the landlord—in the worst case, the antenna and indoor equipment may have to be moved to a different building or location.

- *Roof access and roof rights*—permission to place the antenna on a rooftop and access to that location for installation purposes (including cable runs), determined by existing property leases or other contractual language; may be negotiated into an existing agreement or the subject of a new one.

- *Local zoning restrictions*—government approval of the type of installation and the specific configuration; these may be determined by published building codes, or on the spot by inspectors.

- *Aesthetics*—any need to beautify the external view by obscuring part or all of the antenna, or modifying the physical appearance to comply with either informal or formal demands of local inhabitants.

- *Safety*—measures taken to protect life and property from falling objects or wind-blown debris from the external or internal installation. Also includes safety measures that relate to electrical hazard, a particular concern with high-power amplifiers that require high voltage power supplies and RF radiation.

- *Local weather conditions*—consideration of possible high wind, rainfall, ice and snow, blowing dust and sand, humidity, or other external conditions that could harm the operation of the terminal. It may be possible to relocate the terminal so as to avoid some or all of the difficulty.

Common VSAT network antenna installation options are shown in Figure 9.18. The nonpenetrating roof mount (non-pen mount) is appropriate where the terminal is to be placed on top of a large flat building such as a supermarket or department store. To prevent movement during heavy winds, the truss is weighted down by a ballast consisting of sandbags or concrete blocks. Rocks or gravel may be swept aside to allow a rubberized blanket to be placed between the roof and mount. This provides the friction to

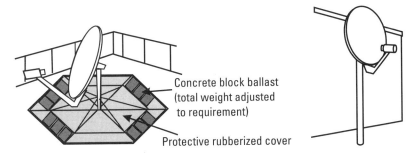

Concrete block ballast
(total weight adjusted
to requirement)

Protective rubberized cover

Non-penetrating roof mount

Ground pole mount

Figure 9.18 Installation options for a VSAT antenna (approx. 1.8m): nonpenetrating roof mount requiring sufficient ballast to prevent motion in high wind; pole mount for ground-level installation.

hold the antenna in place. The pole mount provides a permanent ground mount for convenient access and stability. It may either be attached to the building or the ground itself. Permanent installations often require additional approvals from landlords and local authorities, but may provide superior strength to withstand heavy winds and ice.

The previous discussion is to provide guidance and is not intended as specific installation instructions or authority to use a particular technique. Before proceeding with a particular project, it would be best to have a professional engineer perform a study and arrange for the required clearances.

9.6   Design Requirements for Portable and Handheld Terminals

Portable and handheld communications devices, having dominated the terrestrial mobile radio and cellular scene, are moving into satellite communication. Systems like Inmarsat, ICO, and Globalstar are demonstrating that a lightweight user terminal is both feasible and practical. Applications are developing in remote-site communications, emergency services, security, and even personal convenience while, for example, one is on safari in Africa or mountain climbing in the Himalayas.

To understand the design process for a handheld satellite user terminal we must first examine the architecture of its cellular counterpart (we will refer collectively to all terrestrial mobile radiotelephone systems as cellular). Figure 9.19 provides a simplified block diagram of a generic mobile phone suitable for a digital cellular application (e.g., the GSM standard). Understanding this, we will be able to see how the experience on this side of the industry is being adapted to satellite use (which requires frequency, modulation, and multiple-access changes as well as other changes to the air interface). The RF section at the upper right of the figure contains familiar elements of the RF terminal found in the transmit/receive earth station. The exception is that the modulator is packaged along with the driver and power amplifiers to simplify interfaces within the handheld unit. This reduces RF transmit loss and, consequently, DC power consumption from the battery. Stray radiation within the unit is minimized as well. Both the modulator and receiver (containing the downconverter) may operate on a range of frequencies by way of a common synthesizer. The RF section interfaces with the baseband section through an IF module.

The baseband section of the mobile phone contains both analog and digital sections. We are most concerned with the digital aspects of the unit,

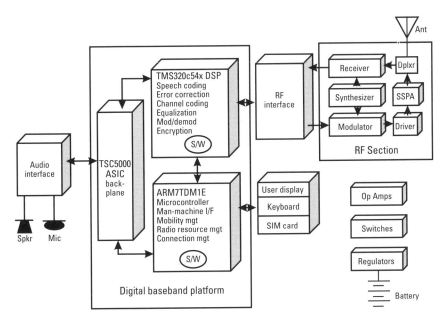

Figure 9.19 Block diagram of generic handheld mobile phone, based on application of ASIC and programmable DSP components.

since these determine the services and operation of the phone as part of a network. In Figure 9.19, all digital baseband processing is performed by digital signal processor (DSP) chips as well as custom application specific integrated circuits. These devices process data under software control, which is introduced through read-only memory (ROM). The latter could be reprogrammed through software download, which is attractive for updating service features and various aspects of the air interface. We see that DSP chips (supplied by Texas Instruments in the example) perform speech encoding/decoding (including compression), forward error correction, channel encoding/decoding (e.g., bit interleaving to minimize the effect of burst errors), demodulation, and encryption.

The functionality of a compact mobile phone is provided and controlled by an onboard microprocessor customized for this application. Functions of the microprocessor include radio resource management (selection of operating frequency and various aspects of channel alignment), connection management (control of off-hook and on-hook conditions), mobility management (handoff between cell sites and roaming features), and human factors related to the control keypad, subscriber interface module (SIM) used within the GSM standard, and the user display. The latter are tending toward small computer screens that switch between normal phone use and other applications such as Internet access and fax display.

The wireless nature of handheld terminal use brings with it many opportunities for variation in service quality and even dropouts. This cannot be overcome within the terminal alone, since it is a function of the uplink and downlink design. The link should stay up with a specified length of dropout. This is one of a number of ways of allowing graceful degradation or a fail-safe form of interruption. As an example, the terminal could anticipate the loss of connection and alert the user to this condition. The user could then alert the other end of the connection that a redial will be required. For networks that extend beyond the reach of one satellite, there is also the need for a roaming feature currently found in cellular networks. A user when arriving in a new service region (e.g., another country), will activate the terminal and wait during the authentication interval. This allows the network to pass messages in the background between the visited region and the subscriber's home network. Once it is verified that the subscriber is valid, the service can automatically be activated.

Several other functions are needed to create a complete handheld mobile phone, whether this is for terrestrial or satellite use. The phone requires a speaker and microphone, which are analog in nature. The amplifiers, noise, and echo cancellation features are contained within the audio interface. Also, there could be provision to connect a PC or other external device. Indicated at the lower right of Figure 9.19 are supporting operational amplifiers (op amps), switches, and DC regulators. The antenna for the phone would be optimized for transmit/receive operation in the desired frequency band. Also of concern is the coverage of the antenna, an aspect that was considered earlier in this chapter.

Dual band and dual mode features are important to reaching the large markets so important to overall system success. A dual band phone is one that works equally in two different frequency ranges, e.g., L band and S band. In the case of terrestrial cellular, dual band refers to the ability of the same unit to function in the 800-to-900-MHz as well as the 1,800-to-1,900-MHz range (e.g., cellular and PCS/PCN). It is assumed in either case that the same standard and air interface are employed. In contrast, dual mode operation means that two different standards may be used, such as:

- Analog cellular and digital cellular (a minimum requirement in the U.S. cellular systems);

- TDMA cellular and GSM in the PCS frequency band;

- GSM cellular and TDMA in the PCS band;

- CDMA cellular and GSM cellular;

- TDMA satellite (ICO) and GSM;

- CDMA satellite (GlobalStar) and CDMA cellular.

Triple mode phones are another outgrowth of the complexity now faced by terrestrial and satellite service providers alike.

An approach to providing two TDMA modes of operation is suggested in Figure 9.20: terrestrial GSM cellular and satellite (L-band) TDMA. We see that the RF section is switchable between the two bands and the baseband section can format it for either of the two access schemes. The functionality can be seamless, meaning that the user cannot tell which mode the phone is in.

This chapter provided an introduction and outline of user terminal design and technology. Because this field is in constant and tremendous flux, it is very difficult to specify a long-term approach to the definition, performance, and manufacture of these important devices. This is because of the rapid evolution of services from satellites and other radio communication systems. Voice and video services, the mainstay of user terminal focus during

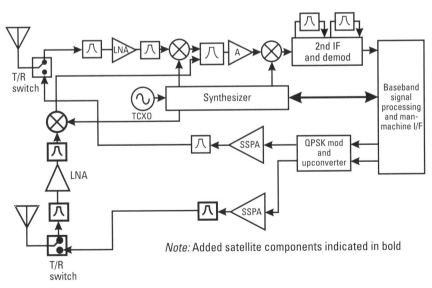

Figure 9.20 Modifications to the design of a standard GSM mobile phone to accommodate a satellite GMPCS mode of operation.

the 1990s, will likely give way to a wide variety of data and multimedia systems [5]. As this same transition is under way in terrestrial networks, one cannot predict how satellite systems will look ten years from now. Technology is also improving, offering greater integration and miniaturization of functions within ASICs and DSPs, bringing functions normally associated with high-speed computer systems down to handheld proportions. While the specifics are impossible to predict, we can be fairly sure that the basic building blocks and RF requirements will remain the same.

References

[1]   Larson, Lawrence E., ed., *RF and Microwave Circuit Design for Wireless Communications*, Norwood, MA: Artech House, 1998.

[2]   Schellenberg, James, "High-Efficiency, Packaged Ka-band MMIC Operating at 24 Volts," *IEEE MTT-S Digest*, 1998, p. 577.

[3]   Yeung, Ping et al. "Design of All-Digital Wireless Spread-Spectrum Modems Using High-Level Synthesis," *Signal Processing in Telecommunications*, Ezio Biglieri and Marco Luise, eds., Proceedings of the 7th International Thyrrhenian Workshop on Digital Communications, Viareggio, Italy, September 10–14, 1995, London: Springer-Verlag, 1996, p. 409.

[4]   Chang, Henry et al., *Surviving the SOC Revolution: A Guide to Platform-Based Design*, Boston: Kluwer, 1999, p. 9.

[5]   Oliphant, Malcolm W., "The Mobile Phone Meets the Internet," *IEEE Spectrum*, August 1999, p. 20.

10

Earth Station Facility Design and Site Selection

Earth stations are like other complex technical facilities such as computer rooms, TV broadcasting sites, and telephone switching centers. All require reliable and complete support from the buildings that house them. Simple user terminals of the VSAT or TVRO class do not require special provisions; however, a major earth station that acts as a broadcast center or telephone gateway needs an integrated system to reliably power and house the electronics, and allow on-site personnel to perform their roles. One key requirement is that heat-generating electronic equipment must be temperature controlled to assure reliable long-life operation. Also, the first installation is usually not the last, as the facility must be modified and upgraded to account for changing requirements. There is also the matter of outdoor equipment, including antennas and RF electronics, which are exposed to the local environment conditions. Facilities of this type also tend to attract attention (sometimes unfavorable) and therefore need to be protected or possibly concealed.

The best way to think of the supporting facility is as the spacecraft vehicle that carries the communications payload of the satellite itself. The space environment that the satellite is exposed to is, in some ways, much more benign than what an earth station in Hong Kong or Siberia, for example, might experience. We are speaking of the following needs, all of which must be continuously met so that reliable services are rendered:

- Prime power, supplied by the commercial power grid;

- Backup power, to supplement prime power during outages;

- Uninterruptible power supply (UPS), to overcome brief outages and prevent damage due to "glitches" and other forms of electrical surges;

- Heating, ventilating and air-conditioning (HVAC), the environmental controls needed to maintain equipment operation, promote long life, and provide a comfortable working environment;

- Building design, layout, and construction, to provide appropriate space for equipment and promote easy access to all aspects of the site by operation and maintenance (O&M) personnel;

- Protection for people and equipment from natural and manmade disruptions and disasters, like fire, flooding, ice and snow, earthquake, unlawful entry, and civil unrest;

- Grounding, shielding, and lightning control, often neglected or inadequately addressed during design and installation phases;

- Protection from RF radiation that can cause human hazard or interference between satellite networks and local microwave systems.

Figure 10.1 provides a system-level block diagram showing the general relationship and interconnection of these elements with the electronic systems. The interesting thing about facility design is that there is no such thing as a standard arrangement. Instead, the facility is laid out according to the custom needs of the earth station and its operators. In this chapter, we review the various systems and some of the more important aspects of their selection and sizing. The specific design will likely be a collaboration of the architects, engineers, and operations team that have a stake in a properly working overall system. The following are some important considerations in this regard [1]:

- Flexibility is key to a successful design—consider current and future needs, taking account of technology that may not exist at initiation of the project.

- Efficiency in design assures that both the initial investment cost as well as the O&M expenses are at their respective minimums. A proper design allows the staff to do their jobs effectively and in an environment that promotes job satisfaction.

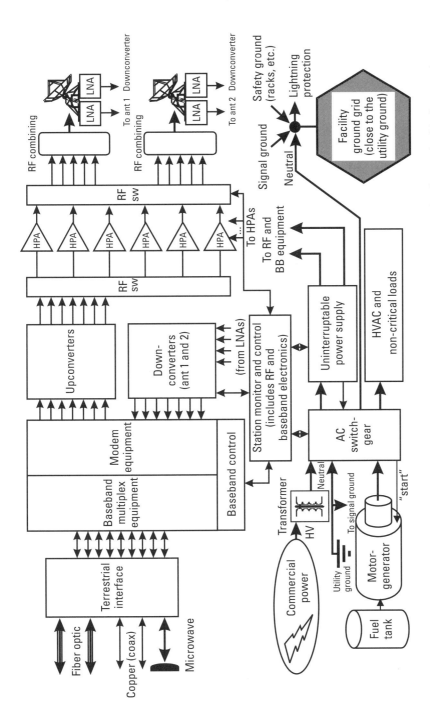

Figure 10.1 General arrangement for a major earth station and its supporting facility, indicating information flow and critical connections.

- Involve in the design process everyone who has a significant stake in the outcome. This includes the architect, engineers, consultants, contractors, equipment suppliers, businesspeople, and the O&M staff who have to live with the result.

- Develop a practical schedule for the implementation of the station and follow the principles of modern project management. The key phases involved are: (1) schematic design, (2) design development, (3) construction documents, (4) request for bid or proposal, (5) contractor selection and negotiation, (6) construction (which may have several phases), and (7) acceptance.

This discussion is an introduction to the general topic of facility design, intended to inform satellite ground segment engineers, who usually arrive at this issue from the standpoint of electronics specification and design, software development, or service operation. These aspects require the expertise of the architect, civil engineer, and electrical power engineer, as well as numerous specialists with years of experience in the issues identified above. Often, the required expertise is obtained from local specialists who understand the building styles, codes, and environments that have an effect on the design and operation of the facility.

10.1   Prime Power and Uninterruptible Power Systems

The single most critical function of an earth station facility is the provision of electrical power to the equipment. The prime power and UPS aspects of earth station design cannot be treated as an afterthought; anyone who has been involved with the installation and operation of an earth station has experienced the difficulties of power failures (in fact, you don't have to be an earth station operator to know the meaning of a power failure). Fortunately, the design of these systems is well understood and there are good technologies to provide high confidence. The key is to understand the nature of the local power system, the needs of the equipment, and practice as to the configuration and O&M of these systems.

According to the National Association of Broadcasters (NAB), many disturbances are beyond the control of the typical commercial power utility company [2], requiring major users to take their own respective corrective action. Insurance underwriters who provide coverage for service interruption have a similar view. The following are some of the types of disturbances that produce undesirable disruptions of electrical power flow. Some only reduce the available current and voltage, while others can either damage equipment or cause it to go offline:

- Large load changes from other customers on a random basis (due to the sudden shutdown or turn-up of another customer's facility). This can produce an undervoltage condition or, for a brief period, an overvoltage condition;

- Power factor (PF) correction switching within the power grid (i.e., automatic controls to correct for lead or lag of current phase with respect to voltage phase);

- Lightning (the most dangerous single type of event);

- Accidental system faults (i.e., loss of input due to a line or transformer failure outside the user's facility).

The amount and type of protection will depend on the particular situation and needs of the facility. A simple VSAT could be protected with a low-cost surge protector, common in personal computers; economical small UPS systems are now available, selling for under $500. A major earth station, on the other hand, will require a customized (and expensive) protection system and, in most cases, a UPS to assure reliable and continuous operation.

A general arrangement of a complete power system for a major earth station is shown in Figure 10.2. The primary purpose of the UPS is to assure continuous and stable power to critical electrical loads. These loads consist of the telecommunications equipment (baseband and RF), computers, interfaces to users, electrical controls, and emergency power for lighting and other functions (such as the break room and the central coffeepot). A motor generator can be added to cover situations where the disruption or outage may last more than the typical 30 minutes that the UPS batteries are designed to cover. One critical need for generator backup is to power air-conditioning to maintain station temperature within an acceptable range. Large earth stations typically require at least one motor generator for these and other reasons.

The following is a list of the types of specifications that a reliable power system would satisfy:

- *Basic power requirements,* including voltage (which is country-dependent—e.g., 115V 60 cycles in the Americas, Japan, Korea, and the Philippines, and 220V 50 cycles in Europe and most of the rest

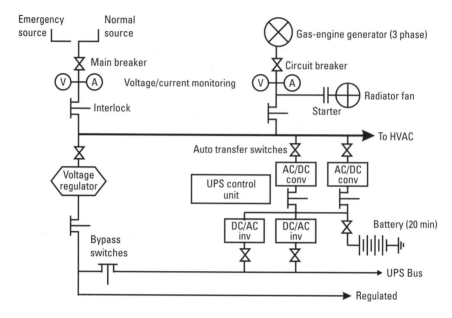

Figure 10.2 Simplified installation diagram for a power system incorporating dual utility feeds, a standby motor generator, and an uninterruptible power supply.

of the world), current, power factor, harmonic content, and transformer configuration. These requirements are determined by the quantity and type of electronic equipment to be installed at the site, initially and at any time in the future.

- *Voltage regulation requirements* of the load. Computer and telecommunications equipment (baseband and RF) are particularly demanding regarding voltage regulation, and so it is normally expected that external regulation by a UPS will be used.

- *Frequency stability and the slew rate*, which is the maximum rate of change of frequency.

- *Effects of unbalanced loading*, e.g., across three phases. This can be balanced within the station itself.

- *Overload and inrush* current capacity and requirement.

- *Bypass capability*, e.g., the inclusion of switching to bypass the UPS or noncritical loads in order to keep the system alive. (In too many cases, this bypass switching system is the single point of potential failure in what otherwise would be a fully redundant system.)

- *Primary-to-secondary-path transfer time* (e.g., the time to switch from UPS battery to backup generators), during which protected equipment will operate on batteries and noncritical loads will be dead.

- *Maximum standby power reserve time*, consisting of the power that the UPS batteries will deliver, as well as the maximum duration of generator operation. The latter will come into question where commercial power could be disrupted for days due to a major storm, natural disasters such as an earthquake, or human errors or actions, including civil unrest.

- *General system reliability and maintainability*, which are requirements driven by a demand to keep the station working under known and unknown situations. One example might be a general "brownout" situation such as existed on the east coast of the United States back in the late 1960s.

- *Operating efficiency.* For example, it might be desirable to run the site on generators to limit commercial power demand (called load shaving) to avoid a penalty fee for exceeding a forecasted load.

The specification of UPS and other load requirements is one of the most important budgeting processes associated with designing the earth station in the first place. This is especially important for large earth stations used as broadcast centers and hubs. Every item of critical and noncritical load must be identified, estimated, and recorded for analysis. Suitable margins must be included, since requirements for newly designed equipment may not be known ahead of time. It is always much better to err on the high side, thus allowing for growth in demand (something that is almost always going to happen). Expanding the basic wiring, UPS, and generator systems after installation and while the station is in operation is a very complex, risky, and costly matter. The more years of smooth and convenient operation you can buy at the beginning, the better for the station during its principal years of service.

The following advice from the NAB should also be taken to heart:

The environment in which the power conditioning system operates will have a significant effect on reliability. Extremes of temperature, altitude, humidity and vibration can be encountered in various applications. Extreme conditions can precipitate premature component failure and unexpected system shutdown. Most power protection equipment is rated for operation from 0°C to 40°C. During a commercial power failure, however, the ambient temperature of the equipment room can easily exceed either value, depending on the exterior temperature. Operating temperature derating typically is required for altitudes in excess of 300 meters (1,000 feet).

10.2   Heating, Ventilating, and Air-Conditioning

Readers who have spent time in earth stations, a major computer center, or broadcast facilities know that the internal temperature is usually on the low side (around 20°C or 68°F), uncomfortable unless you are wearing a long-sleeve shirt or a jacket. This standard requirement is designed to allow the electronic equipment to function normally, relying on the conditioned air to control temperature. In addition, the HVAC systems keep the humidity within a safe range for equipment, lest it be exposed to more moisture than manufacturers allow. Because these factors are critical to safe functioning of technical equipment, the performance of the HVAC system must be stable and reliable, often requiring a fully redundant design.

The preferred means of providing conditioned air to racks of equipment is through the bottom, allowing the cooler air to replace warm air that exits from the top. The air reaches the rack underneath the flooring or through special ducts. Normal practice in major earth stations and computer centers is to install equipment on top of raised computer flooring. This has the additional advantage that the flooring panels, which are supported by a structural framework, can be removed to allow access underneath the racks. This space is also employed to route the cables within the station, which simplifies the job for technicians and engineers who install and upgrade the systems. The simplest approach, however, is to allow the equipment racks or units to take in the air directly from the room itself. Regardless of the approach, the HVAC system will be required to maintain temperatures in the required range. One cautionary note: water used for humidification should be filtered to remove chemicals, particularly chlorine, to prevent damage to sensitive electronic parts.

The *H* in HVAC stands for heating; however, it usually turns out that the electronic equipment itself provides sufficient waste heat to keep the temperature from falling too low. This may not always be the case, particularly in very low temperature regions during winter. Most facility designs focus on the air-conditioning aspect of the HVAC system. The key, then, is to have high confidence that it is working at all times. In an unattended station, a good monitoring and alarm system will provide early warning of HVAC difficulty. Immediate attention to these types of problems will ensure that a major station outage, with ensuing equipment damage, does not occur.

10.3 Building Design and Construction

The building that houses an earth station or other major ground segment facility is very much a part of the entire system. If the service and technical requirements are determined properly in the first place, the building will likely come together in an effective way and meet these requirements. Costs can then be estimated with reasonable accuracy and the project can follow its planned schedule, resulting in a completed facility that is on time and within budget. Organizing this overall effort is usually the responsibility of the facilities engineering manager, who supervises the architectural engineering team and the construction manager; they work as a team to collect requirements, design the building to accommodate the mission, equipment, and people, and oversee the construction to a proper conclusion. Building design does not end when the building is complete. This is because there is always a need to add capabilities to the building to accommodate changes in requirements. In fact, the lifetime of a major earth station facility can be substantially longer than that of the specific baseband and RF electronics that are providing specific services at any given time.

A new building is a major challenge for an organization, complicated by the number of people involved in the requirements and the task of properly supporting the technical equipment and its operation. Earth station sites are often located in isolated places where ground access may be restricted. Site selection and preparation is often a lengthy and expensive process. Any given site must have strong benefits in terms of physical space and satisfaction of the technical requirements of orbit visibility and isolation from RF interference. Unfortunately, what might be an ideal site for technical reasons could be a nightmare for the construction crew or the operators who have to work there on a round-the-clock basis. All of these factors have to be considered in advance, before the site is selected and construction begins. Turning a project around after it is under way is extremely hard, and the O&M staff may suffer (and complain) for years to come as a result.

The task of the facilities engineering manager is often as difficult as that of communications project managers who put the satellite network together. Often, this person starts to work during system development and construction, and remains with the operating facility in an O&M role. In this way, the particular individual can help with new requirements that come along from time to time, ensuring that operations continue smoothly as the changes are implemented. Site upgrades are often more difficult than the original design and construction task because the work must be performed on a revenue-producing station. Some of the responsibilities of this position include the following [3]:

- Collecting requirements for and planning of the facilities that will house the earth station and the personnel who will operate and maintain it (including the building and its systems);
- Converting these requirements into technical specifications that can be used to engineer and design the facility;
- Engineering and design of the building and its systems;
- Engineering and design of modifications to the building and its systems, required from time to time after the earth station goes into operation;
- Engineering and design of maintenance and repair systems, all the more necessary if the site is in an isolated location;
- Construction of the facilities and installation of the earth station equipment;
- Maintenance and repair of facilities and equipment;
- Generation, purchase, and maintenance of utilities systems;
- Cost management, to evaluate ways to reduce costs (fixed and variable), possibly by replacing obsolete equipment to save energy costs or to improve operational efficiency.

We can see from this summary that the tasks cover a very broad range of time and activity, starting when the project is conceived and lasting to the end of the useful life of the earth station. This is a lot like building your own home and then having the pleasure of living in it for many years. Much of the effort has to do with identifying and controlling costs: initial costs (investment) and recurring annual operating costs. The following list provides the categories of typical costs that are encountered in facilities of this type.

- *Site development costs:*
  - Survey and license expenses;
  - Purchase of land;

- Access roads;
- Clearing, grading, and preparing land;
- Extension of water, sewer, commercial power, and telecommunications (fiber-optic or microwave);
- Fences, gates, guard station, TV monitors, alarms, and other security measures;
- Internal roads, walkways, overhead covers, and parking areas;
- Exterior lighting;
- Landscaping.

- *Construction costs:*
  - Foundations for buildings, utility connections, antennas, and cable trays or trenches;
  - Buildings (new construction or renovation);
  - Utility distribution systems;
  - Storage yards for materials and other resources;
  - Power generation systems;
  - Antennas and towers;
  - Special facilities for storage and testing (if required);
  - Storage tanks;
  - Waste processing (if required);
  - Site RF shielding (if required).

- *Operating costs:*
  - Utilities usage (electrical, water, gas, etc.);
  - Telecommunications costs (telephone, wideband connections [e.g., fiber optic], wireless/cellular, LAN and WAN, etc.);
  - Maintenance and repair of facilities;
  - Operation of power generation and other facilities equipment;
  - Custodial services (e.g., cleaning);
  - Operation and maintenance of automobiles, snow-removal equipment (if appropriate), and other vehicles;
  - Trash collection and disposal;
  - Inspections and corrective actions;
  - Fire and police protection;
  - Insurance, licenses, and other routine fees.

- *Support equipment costs* (fixed items that must be replaced from time to time):
  - Office furnishings, lab benches, and miscellaneous installed items;
  - Automobiles and other support equipment for the site;
  - Electrical and communications cabling;
  - Additional licenses and permits as required;
  - Moving costs;
  - Lost productivity (opportunity) costs.

Earth station buildings differ from typical buildings used for manufacturing or R&D. Like the spacecraft they use as a relay, earth stations are laid out for the transmission function they perform. Below is a list of the flows that need to be considered.

- *Signal transfer and routing:*
  - Transfer of user information from the terrestrial interface to the baseband equipment that initiates the transmission function of the facility;
  - Routing of baseband signals between racks of equipment and computer systems, as appropriate;
  - IF and RF transmission within the building, leading to the HPAs and LNAs on the uplink and downlink, respectively;
  - Transmission between antennas and the RF equipment, accomplished by the interfacility link (IFL);
  - Reference frequency and/or station time;
  - Intercom.

- *Acquisition and distribution of services:*
  - Power distribution, between commercial sources and backup systems;
  - Distribution of reliable power from the UPS to critical electronics;
  - HVAC distribution to maintain temperature and humidity;
  - Water for various purposes, such as sanitation, safety, humidification, and cooling;
  - Dry air or nitrogen for waveguide purging.

- *Efficient flow for staff* during the care and operation of equipment and the building systems themselves. Stations located in extreme climates pose a special problem to staff who will, at times, have to go outside into the elements to reach equipment and parts of the facility needing attention.

- *Access to electronics for O&M activities*, including repair and replacement of units and components, and modifications and upgrades from time to time.

- *Providing a safe environment for staff*, no matter what their task (during all phases of construction, operation, and future modification).

- *Providing a comfortable temperature range* (18° to 22° C), sufficient lighting, and tolerable levels of noise for an extended period.

Even though every earth station uses similar elements and facilities, there is no such thing as a standard facility design. Every design effort must begin at the beginning, addressing the special requirements for the particular system and physical situation. Large facilities that combine an earth station with a telecommunications gateway or TV broadcast operation center must address the needed services to support the added staff and functions. For example, a typical earth station providing a few uplinks and downlinks can probably be operated by a team of six; however, expanding the function to include video origination for a DTH network would see staff requirements grow to nearly 100. It is best to work closely with the architects and specialists who can interact with the designers and operators of the system. It is also best to allow for changes during the project, and to maintain a constructive attitude toward new ideas and questions. We have covered many of the issues and standard needs for design of the facility; however, this cannot replace a close collaboration among the interested parties, particularly those who have to live and work within the earth station once it is completed.

10.4 Grounding and Lightning Control

The electrical power and communications signals within an earth station flow through a complex network of physical/electrical connections that are supposed to allow for a smooth operation of systems and the building itself. The electrical codes and other rules for wiring dictate proper practice, which when followed provides for safe operation under most conditions. However, an earth station is much more complex than the standard industrial building in terms of electrical and electromagnetic energy transfer. For example, we have high-power energy in some cases competing with very low level electrical and microwave signals that must pass from rack to rack undisturbed. Another factor is the possibility of low-quality commercial electrical power and the need for backup, coupled with danger from lightning strikes on the electrical lines or to local metal structures such as antennas, towers, and the building itself. What you end up with is an overall environment that can contribute to system disruption and possibly a reduction to the safety of on-site staff. Many of these problems, even if anticipated, do not show up until the facility is completed and the first operation is attempted (or possibly months later). The corrective action may require a considerable expenditure of money and even take part or all of the earth station out of service. Of course, both of these outcomes are better than a situation in which people are at risk. The following paragraphs provide some guidelines for addressing the basic issues of grounding and lightning protection to provide an appreciation for the need for proper design and verification. However, this is no substitute for a detailed investigation and testing program at appropriate phases of building construction and while in service. Part of the plan should include checking and following local electrical codes and other relevant regulations, which can differ from country to country, state to state, and even city to city.

10.4.1 Principles of Grounding and Noise Control

According to the IEEE, the term *grounding* refers to a conducting connection, whether intentional or accidental, that causes an electric circuit to be connected to earth or to a conducting body of large extent that serves the same purpose in electrical terms [4]. To be a ground, the earth (which may be augmented with a wire grid to improve conductivity) or large conducting body (such as equipment racks or a metal structure of some type) must establish a constant potential to allow conductors to be connected to it. There are at least four reasons for this ground:

- To provide personnel safety by limiting the electrical voltage difference between metal parts around the station and the ground itself;

- To prevent or control ground loops, which are unintended paths for noise and stray signals to find their way into sensitive receiving and data processing elements of the earth station;

- To control electrostatic discharge (ESD), which will damage sensitive electronic components and cause unpleasant human shocks, by limiting voltage difference between noncurrent-carrying metal elements;

- To provide fault isolation and equipment safety by giving a low impedance fault return path to the power source.

Noise control is very important in earth stations and computer facilities. Ideally, the ground should be at the same potential throughout the station, and racks and cables should be provided with excellent shielding to resist the radiated transfer of noise and other undesired signals. However, there is always a voltage difference across cables and between racks; if there are separate rooms and even buildings making up the station, the situation is made even worse. Where the physical separation is large enough, there could be a requirement to isolate one section from the other. It is preferable to connect rooms and buildings using fiber-optic cabling, because such cabling will not, as a rule, allow conducted noise to be transferred. This is one excellent application of fiber optics within the earth station facility.

10.4.2 Lightning Protection

Lightning is one of the biggest concerns in earth station safety, both for the people and equipment, particularly in tropical and semitropical regions where thunderstorms are prevalent. Electronic equipment that is interconnected by wire cables and provided electrical power from commercial sources is vulnerable to damage or destruction from lightning-induced electrical transients [5]. As shown in Figure 10.3, a direct strike to the building, metal structures (towers and antennas), commercial power lines, or the ground itself can pose a threat. In some cases, the lightning may not even hit the earth but involve flashover, electromagnetic induction caused by cloud-to-cloud discharge, or shift of ground potential caused by currents from a strike dissipating through the ground itself. According to Murphy's Law, some of this energy will find its way to—and possibly damage—the most sensitive equipment (RF electronics, baseband and computer systems, UPS components, and computing devices). Also, the power supplies used to convert AC to DC throughout the station are always the first point of entry of lightning-induced current surges. The failure can come through a subtle mechanism involving transistor junctions within a specific circuit, even though the actual lightning hit might occur many kilometers away.

There are two major issues regarding lightning protection: (1) designing the building and equipment installation so that the majority of lightning

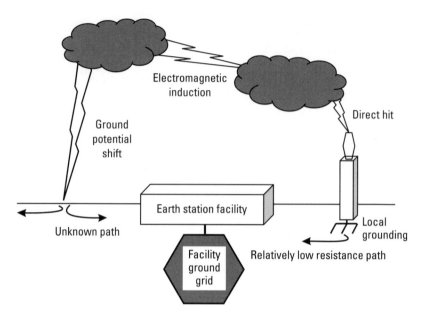

Figure 10.3 Direct and indirect paths for lightning to reach the interior of an earth station.

hits do not cause disruption or harm, and (2) taking measures to reduce the incidence and intensity of these hits in the first place. The first point involves taking all the right steps in providing grounding and appropriate isolation, while the second is much more difficult to achieve in practice. We use various forms of surge protection to provide a nondestructive path for surge current to flow to ground, rather than through the equipment. At the same time, the protector will limit the voltage across the protected component to again improve ground electrical flow. Surges can occur in "common mode," meaning between a set of signal wires and ground, and "difference mode," meaning across a pair of wires in a signal path. The common and difference modes are also how noise and spurious signals are conducted between racks of equipment and buildings (fiber-optic cables are a good cure for conducted noise).

Lightning-induced surges can be rather broadband, meaning that they contain high frequency energy. Surge protection must have a low inductance path to earth so that the surge does not produce a momentary voltage transient (and hence damaging current flow). This suggests that surge protectors are appropriate at both ends of every long cable run. Note also that metal conductors of coaxial cables and waveguides offer pathways for surges and noise.

The topic of lightning protection is perhaps the most complex and least well understood of all the critical areas of station facility design. The techniques are all specific to the particular situation, e.g., a good solution at one facility in one particular location may be totally ineffective for another application somewhere else. Of course, old standbys of lightning rods and proper earth grounds will be the standard approaches. But when the first actual lightning strike occurs, the outcome is never easily predicted. Stations that have received damaging strikes must have the attention they deserve. The basic grounding systems are always suspected when outages occur, and the necessary corrective action probably won't be quick and easy. In one particular case in the tropics, the process lasted more than a year and involved a series of particular improvements. Among these were: converting from coax to fiber to connect buildings within the same site; installing more ground straps and buried wires; greater use of surge protection at equipment racks; and providing improved lightning rods on nearby structures.

10.5   Radio Frequency Clearance and Interference Analysis

It is a fundamental principle of radio communications that other users and services share the frequencies we employ. Thus, there is no such thing on this planet as a dedicated sole-user frequency assignment—any frequency is subject to occupancy by more than one independent transmitter. Of course, there are many factors that allow us to effectively reuse frequencies without mutual interference. The techniques that we apply fall under the broad category of radio frequency clearance; if clearance is not guaranteed, then we embark on frequency coordination, a potentially laborious process of interference analysis, test, and negotiation. Specifically, frequency coordination is a process of discussion, investigation, and negotiation among the affected parties, under the general principle of first come, first served. In other words, whoever is first to use a frequency—provided they have completed all of the required regulatory processes to achieve a protected position—has some domain over the use of that frequency. The degree of domain and control that the first protected user has depends on the particular part of the spectrum, the rules that apply under the International Telecommunication Union (ITU), and the local government rules and regulations in the country of interest. There are exceptions to the first-come-first-served principle, primarily in cases where the ITU has devised an a priori plan for that part of the spectrum (e.g., Broadcasting Satellite Service at Ku band).

10.5.1 Role of the ITU in Frequency Coordination

Founded in 1865, the ITU is the oldest and most prominent of international organizations in the field of telecommunication regulation. As a specialized agency of the United Nations and with its origins dating back before the age of radio, the ITU has proven itself in a changing and turbulent world. Today, it has over 150 member nations, with its headquarters in Geneva, Switzerland. Membership is restricted to government telecommunication agencies, called administrations; each is required to contribute to the Union's financial support more or less in proportion to its economic strength. Telecommunication service providers, satellite and earth station operators, and users are represented through their respective governments and often attend meetings along with their government counterparts.

The objectives of the ITU, reflecting the shared purposes of the administrations, are:

1. To maintain and extend international cooperation for the improvement and rational use of telecommunications of all kinds;
2. To promote development of technical facilities and their efficient operation, so as to improve telecommunications services and increase their usefulness and availability;
3. To harmonize the actions of nations in attainment of these goals.

Generally speaking, the ITU does not have direct power to regulate radio transmission, as this is considered to be the right of sovereign states. However, because the ITU is a common body among governments, its regulations and recommendations are followed by the member nations out of the practical need for a realistic framework. There have been instances where politics has entered into the process when particular interests see the opportunity to work as a block and pursue parochial goals. On the other hand, developing countries have a special place in the ITU because of the shared objective of improving worldwide telecommunication services. A country with a poor telecommunication infrastructure cannot provide international telecommunication services of good quality. Furthermore, there is the more general UN mission of improving the living conditions of the world's population, for which the ITU carries out the telecommunication aid program.

A principle area for the ITU is management of the radio frequency spectrum, a critical role because individual countries and operators of radio transmitters need "rules of the road." Otherwise, radio stations would transmit on any frequency they choose and interference would be very common;

therefore, no one would be able to count on reliable radio broadcasting and point-to-point radio communication. Through international conferences and the Radio Regulations, the ITU oversees the use of radio frequencies and provides an effective forum for the resolution of interference difficulties. However, it does not actually police the airwaves.

The Radio Regulations of the ITU (also called the Red Books, due to their color) contain the rules and procedures for the planning and use of all radio frequencies by administrations, who in turn assign individual frequencies to their respective government and private earth station operators. Unlike legal codes, the Radio Regulations are usually not drafted by lawyers but rather by experts in frequency management and telecommunication. A key section shows the allocations of frequency bands to particular services, such as the Broadcasting Satellite Service (S and Ku bands), the Fixed Satellite Service (C, X, Ku, and Ka bands), and the Mobile Satellite Service (L and S bands).

In addition to being a horizontal division of the total spectrum by service, there is a vertical division by geographical region: Region 1 includes Europe and Africa, Region 2 North and South America, and Region 3 Asia and the Pacific.[1] The Table of Frequency Allocations, which presents the division between services and regions, also contains dozens of footnotes allowing administrations to make exceptions to the rules. For example, a footnote may permit country A in Region 1 to assign a terrestrial microwave radio user to a frequency in a microwave band allocated exclusively to the Mobile Satellite Service.

The ideal procedure for an administration to follow is to assign frequencies only in accordance with the Table of Allocations and Rules of Frequency Coordination. If the transmission can potentially cause interference, then the administration must follow the convoluted process of frequency coordination, which is also defined in the Radio Regulations. This gives the other administrations a chance to decide if they want to allow this particular station to go on the air. A successfully coordinated frequency assignment can be recorded in the Master International Frequency Register (the "Master Register") of the ITU and thereby gain international status.

What follows is a brief overview of the Radio Regulations; it should not be used in an actual application.

There are basically two types of frequency coordination: terrestrial coordination, for earth stations and land-based microwave transmitters, and

---

1. Ray Sperber suggests the following mnemonic: Region onE (Europe, where coordination started), Region tWo (the New World of the West), Region thrEE (EastErn Asia).

space coordination, for radio transmitters and receivers on satellites. Terrestrial coordination (discussed in Section 10.5.2) involves any terrestrial radio transmitter that can potentially radiate signal power across a border into a neighboring country. Included are earth stations and other types of terrestrial microwave transmitters with sufficient power to cause transborder radio frequency interference (RFI).

The process for coordinating a new satellite can be very lengthy because of the potentially wide radiation pattern across the earth and because more frequency bandwidth is usually involved. Coordination is required with other satellite networks that fall into one of the following three categories:

- Those that have been recorded in the Master Register of the ITU-R;
- Those that have already completed coordination;
- Those that have entered coordination before the new network in question.

Coordination of satellite networks is a bilateral activity in which the newcomer must approach the incumbent and obtain their agreement regarding the potential for interference between systems. Such discussions and negotiations can take months or even a year or more in particularly difficult or acrimonious situations. This difficult process could be helped by a new arrangement under the WARC 88 Final Acts, wherein a multilateral planning meeting (MPM) can be called by an administration.

Most of the time, coordination and registration are accomplished in a straightforward manner, taking anywhere from six months to three years, depending on the number of administrations involved and the complexity of the technical analysis of potential interference. Occasionally, an administration locks its heels and refuses to allow coordination of a new frequency assignment by another administration. In such a case, the ITU can be brought in to mediate.

Because the previous discussion was a brief summary, readers wishing more background can consult the very informative Web site of the ITU, www.itu.int. Our focus from here on is the RFI mechanisms that affect earth stations and the services they render, and the general process that engineers go through to determine if a site is acceptable and can be coordinated.

10.5.2 Terrestrial Interference and Coordination

The location of an earth station can be determined by many factors, such as closeness to a company headquarters or business accommodations, availability of utilities or other services, presence of a good community where the staff and their families can live, and connection to existing telecommunication facilities, roads, and the like. These are certainly important factors, but our focus here is on the control of RFI that the earth station may either cause or receive. In the first generation of C-band earth stations for INTELSAT, the same frequency band was in use extensively around the world for terrestrial line-of-sight microwave radio links. Often operated by the same communications service provider as the earth station, siting was a definite challenge since microwave radio was used for the space link as well as the terrestrial tail connection back to the telephone exchange or TV station. These considerations produced earth station sites that were very far from cities, needing long-haul microwave or cable tails that would not cause RFI into the earth station (and vice versa).

In time, C band became more tame to use, as much because the frequencies are more applied to satellite services as because the satellites themselves transmit with higher power and therefore can overcome the RFI posed by terrestrial microwave systems. We now have C-band earth stations that operate close to downtown districts of major cities like New York, London, Beijing, Hong Kong, and Rio de Janeiro. In many cases, the natural RF shielding provided by the curvature of the earth, hills, trees, and construction must be enhanced through artificial techniques like concrete and metal walls, metallic mesh screening, or containment of the antenna in a pit. However, the fact that C-band-transmit earth stations have a potential to interfere with terrestrial microwave links (even those hundreds of kilometers away) places on them a greater burden of frequency coordination.

Adoption of Ku band offered for the first time the potential to place earth stations literally anywhere the operator or user desired. Thus began a revolution in the use of satellite communications, where TV receive-only terminals and two-way voice and data communications earth stations began to spring up in cities and suburbs as well as on roving vehicles and even aircraft. The modern VSAT exemplifies the kind of flexible earth station that can be operated from almost anywhere. The FCC of the United States even adopted simplified rules, called blanket licensing, that allowed developers of VSAT and later DTH at Ku band to create large-scale networks with a single license.

Many principles first introduced in the United States by the FCC are being adopted in Europe, Latin America, and Asia. Many countries in Latin America follow blanket licensing at Ku band. However, at the time of this writing, Ku-band blanket licensing was still not an accepted principle in Europe, nor has it been accepted by the ITU. C band still must be coordinated with terrestrial microwave systems, an often laborious and time-consuming process. Yet C band still offers the advantage of less fading due to rain, making it more suitable for high-availability applications in rainy climates—particularly in the tropics.

10.5.3 Interference Entries

RFI that is on the same frequency (cofrequency) as our desired signal, or overlaps our bandwidth, is potentially a disruption to communication. The other side of the coin applies as well: our signal can potentially interfere with other channels of communication. Listed below are four levels of impact from RFI:

1. Interference that cannot be detected (and hence does not exist for all intents and purposes);
2. Interference that can be detected but that does not cause unacceptable disruption of another service (e.g., acceptable interference);
3. Interference that causes unacceptable disruption (e.g., interference that, while disruptive, would not render the service totally unusable);
4. Harmful interference, which is interference that prevents usable communications for the interfered-with service.

While it is difficult to provide general rules of engagement for coordination, we can offer some typical examples of how the four levels of interference may arise in the real world. These are suggested in Table 10.1.

There are a number of standard interference paths, called interference mechanisms, that we encounter in satellite and terrestrial microwave communications (illustrated in Figure 10.4). These are controlled through various regulatory mechanisms, including frequency coordination, as will be discussed later in this chapter. Normal operation of the respective links is indicated by the solid arrows, which provide full duplex transmission. The terrestrial link in this illustration operates in both the satellite's uplink band (assumed to be 6 GHz) and simultaneously in the downlink band (assumed to be 4 GHz). Other bands, such as 11 and 18 GHz, which are shared between terrestrial and satellite, would have similar interference paths. The interference paths, indicated by the dashed arrows, can be described as follows:

- *Path 1*—6 GHz terrestrial radio interference into the uplink receiver of the satellite. The satellite is protected by a maximum radiated

Table 10.1
How Interference Affects Satellite Communication Services

Level of Interference	Impact on Analog TV	Impact on Digital TV	Impact on Data
Level-1 (RFI cannot be detected)	None	None	None
Level-2 (RFI can be detected but effect is acceptable)	Slight decrease in S/N, with no perceptible impact on picture quality	None, unless the signal is close to threshold	Increase in error rate that is less than specified %
Level-3 (RFI causes unacceptable interference)	Visible snow and possible disruption of viewing quality	Causes threshold to occur at nominal signal levels	Significant increase in error rate, such as by order of magnitude
Level-4 (RFI causes harmful interference)	Picture rolls or breaks up; unwatchable	Receiver cannot acquire digital stream	Data reception is unusable

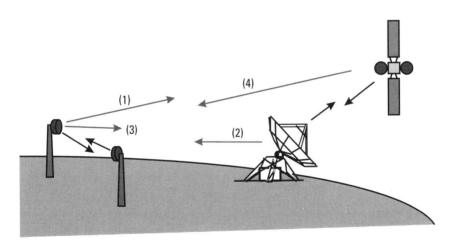

Figure 10.4 Interference potential between satellite links and terrestrial microwave links that use the same frequency band. Paths 2 and 3 are of direct concern in terrestrial coordination; paths 1 and 4 are covered in the radio regulations of the ITU.

power limit of the terrestrial microwave antenna and by a stipulation that these antennas should not be directed at the GEO. As a result of the low radiated power from terrestrial stations, interference of this type is not experienced in practice. The one possible exception is high-power tropospheric scatter links that employ billboard-sized antennas. These must not be pointed at the GEO in any instance.

- *Path 2*—6 GHz earth station interference into the 6 GHz terrestrial radio receiver. This is the most important interference case for determining if a transmitting earth station can be operated in a particular location. The radiation along Path 2 is in the sidelobes of the earth station antenna and propagates along a variety of paths to the terrestrial receiver. The national government in a given country asserts that new transmitting earth stations validate that their operation will not cause unacceptable (or harmful) interference to the operation of existing or planned terrestrial microwave stations in proximity to the proposed earth station site. The ITU has regulations and procedures to coordinate earth stations in one country that could interfere with terrestrial microwave links in adjacent countries.

- *Path 3*—4 GHz terrestrial radio interference into the downlink receiver of the earth station. A common term for this is terrestrial interference (TI) and has been one of the reasons why DTH services use Ku. The earth station antenna could be shielded from the terrestrial transmitter, a technique that has been applied to Path 2 as well.

- *Path 4*—4 GHz satellite interference into the 4-GHz terrestrial radio receiver. The radiation level from a satellite is typically too weak to be of much concern to terrestrial receivers. Still, the ITU places a maximum allowed power flux density (PFD) from the satellite on the surface of the earth.

Path 2 is of concern for transmitting earth stations while Path 3 must be addressed for any satellite receiving system on the ground. Most Ku-band receiving earth stations are protected from Path 3 TI because the band is not shared with terrestrial microwave (except in some exceptions). Operation of user terminals at L and S band should likewise not experience either Path 2 or 3, but there are a number of instances where exceptions were granted for particular countries. As a result, any new operation at these frequencies should be evaluated ahead of time to assure the system designer that terrestrial stations are currently not in operation.

10.5.4 Analysis Methods

Interference to satellite communication ground stations as well as to terrestrial stations that share the same frequency bands can be evaluated using the familiar technique of carrier-to-interference (C/I) calculation technique [6]. For the receive earth station geometry illustrated in Figure 10.5, the C/I equation can be expressed in dB as follows:

$$C / I = [G(0) - G(\theta)] + [P_t - P'_t] + G_s - G'] - 20\log(R / R')$$

where $G(0)$ is the gain of the earth station antenna in the direction of the satellite; $G(\theta)$ is the gain of the earth station antenna in the direction of the terrestrial station; P_t is the transmit power of the satellite into its antenna; P'_t is the transmit power of the terrestrial station into its antenna; G_s is the gain of the satellite antenna in the direction of the earth station; G' is the gain of the terrestrial antenna in the direction of the earth station; and R and R' are the range between the earth station and the satellite and terrestrial station, respectively.

This equation shows the terms that are differences (in dB) between the desired and interfering contributors. Perhaps a simpler way to examine the problem is in terms of an interference budget, which is nothing more than two link budgets: one for the desired path and the other for the undesired (interfering) path. The C/I is the difference, in dB, between the desired and interfering received carrier powers. An example of an interference budget is presented in Table 10.2 for the interference geometry indicated in Figure 10.6.

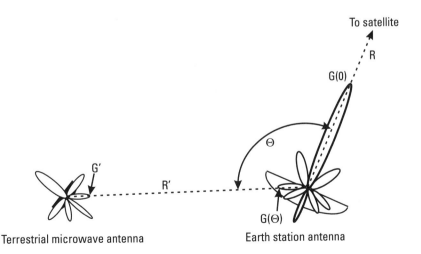

Figure 10.5 Interference geometry (top view) for terrestrial interference to reception at an earth station with a large-diameter antenna.

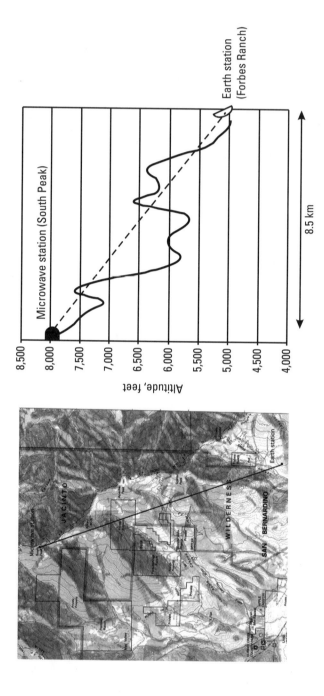

Figure 10.6 Example of an interference path study for a mountaintop microwave station and a well-shielded earth station antenna.

Table 10.2
RF Interference Budget for a C-band Digital Video Link Interfered by a Terrestrial Microwave
Transmitter Operating on the Same Frequency Channel

Factor	Value	Units
Desired signal (transmitted by satellite)		
Satellite transmit power per carrier at antenna port	20	Watts
Satellite antenna gain in direction of earth station	30.0	dBi
Satellite EIRP in direction of earth station	43.0	dBw
Free space loss	196.2	dB
Atmospheric loss (no rain)	0.1	dB
Earth station antenna gain (3m dish)	39.8	dBi
Received desired carrier power	−113.5	dBW
Undesired signal (transmitted by terrestrial microwave station)		
Transmitter power per carrier at antenna port	10	W
Transmit antenna gain in direction of earth station	0.0	dBi
Terrestrial microwave EIRP in direction of earth station	10.0	dBW
Free space loss (8.5 km path)	132.1	dB
Terrain shielding	20	dB
Earth station antenna gain in direction of terrestrial microwave station	0	dBi
Received interference power	−142.1	dBw
Carrier-to-interference ratio (C/I)	28.6	dB

There are several significant issues that come about when evaluating C/I in this manner. Some relate to the actual physical characteristics being modeled (e.g., the sidelobe gain of the earth station and terrestrial microwave antennas and the amount of expected terrain shielding or blockage to be realized), while others are statistical in nature (e.g., atmospheric absorption and tropospheric scatter propagation of the terrestrial microwave signal). Another extremely important aspect of the study is how the interference actually affects the desired signal.

The analysis in Table 10.2 is straightforward and makes certain assumptions about the strength of the interference. The terrestrial station is 8.5 km from the earth station and the path is partially obstructed (e.g., with 20 dB of shielding from hills). At a C/I of 28.6 dB, the interference would be detectable and may affect reception at the earth station. The degree of degradation will be determined by the following factors:

- *The center frequency of the interference relative to the desired carrier.* The worst-case assumption would be cofrequency (e.g., both carriers centered on the same frequency).

- *The bandwidth of the interference relative to the desired carrier.* The worst case will depend on the type of modulation used on each. For PSK on both carriers, the worst case is equal bandwidths, while for FM the worst case is fully modulated on the desired carrier and unmodulated interference.

- *Whether the interference has constant modulation or time-varying modulation.* The worst case depends on the factors in the previous case, coupled with the time occupancy of the interference within the bandwidth of the desired signal. Analysis of this type of interference situation is quite complex and may require computer simulation and possibly laboratory measurement.

Interference criteria in terms of C/I are determined ahead of time using commonly agreed standards as well as values set during coordination. Table 10.3 provides examples taken from ITU-R recommendations. Some entries contain a single value that provides assurance of acceptable interference (e.g., neither harmful nor unacceptable, but could be measurable). Alternatively, a formula takes account of the difference in bandwidth of the interfering signal as compared with the desired signal. The term B/b is simply

Table 10.3
Examples of Recommended C/I Values for Use in Coordination

Type of Desired Carrier	Type of Interfering Carrier	Recommended C/I (dB)	Comment
Analog TV	Analog TV	17-29	ITR-R Rec 792
Digital TV	Digital TV	23-30	ITU-R Rec BO.1297
Narrowband digital data (SCPC)	Analog TV	$20 + 13\log(B/b)$	Estimated
Wideband digital data (TDMA)	Analog TV	21	Estimated
Narrowband digital data (SCPC)	Wideband digital data	$21 + 10\log(B/b)$	Estimated
Wideband digital data (TDMA)	Wideband digital data	21	Estimated

the true ratio of the bandwidth of the interference to that of the desired signal (assumed to be less). In the case of wideband digital interfering with narrowband digital data, this is the fraction of interfering carrier power that falls on top of the desired carrier bandwidth. For an analog FM video signal, the interfering carrier component is swept in time at a rate of approximately 15 kHz across the bandwidth of the desired narrowband data signal. Potentially unacceptable or harmful interference is possible during the fraction of time that the interfering carrier is physically within the bandwidth of the desired carrier. A general practice where two standard QPSK carriers are interfering with each other is to treat the interference as an equivalent amount of thermal noise.

Propagation losses, other than free space loss, may represent a significant challenge for anyone trying to put a realistic interference budget together before field measurements can be made. A process for doing this is suggested in the next section. Some of the other factors relating to earth station and terrestrial antenna performance, particularly in the backlobe and sidelobe regions, are also difficult to assess. Sometimes these become the focus of dispute and eventual negotiation between the operators who are party to the coordination process. Values of C/I likewise become part of the negotiation, since they determine how the interference is to be considered in the first place. The newcomer, seeking approval from the incumbent operator, would tend to accept more interference than the latter. Just how many dB there are between the two parties will often depend on factors outside the technical aspect of the negotiation.

10.6 Site Selection

The selection of an earth station site, particularly one that operates in a frequency band that is shared with terrestrial microwave radio systems, can require a significant expenditure of time and money. Our objective is to meet all of the technical requirements for the station and to minimize the possibility of receiving or causing unacceptable interference.

The steps in RFI evaluation for site selection include the following:

1. Performing a preliminary survey of existing microwave stations that operate in the same frequency band. This involves researching the records of past frequency assignments in the particular area and locating those transmitters and/or receivers on a map of sufficient scale to indicate all potential problems. This considers all stations

that operate on the same frequency channel as well as adjacent channels that allow for spectral overlap.

2. Creating a topological map representation of the potentially interfering and interfered-with stations (including the earth station itself), and projecting the radiation beams to show how they would intersect. This is done without regard to local site shielding from hills, trees, and buildings (these factors are brought into the picture later, if their presence is required to meet the C/I criterion). The antenna or antennas of the earth station would be drawn with their respective beams pointing toward the satellite or satellites that might be employed. An example of such a map is shown in Figure 10.6. Terrestrial stations whose antennas would not pose a problem can be removed from the study (this process is said to "cull" terrestrial stations that can be ignored in the site selection process).

3. A detailed evaluation of path profiles is then performed for the terrestrial stations that could be exposed to RFI threat (and vice versa). This will resolve which paths could produce RFI, taking account of the natural shielding afforded by the curvature of the earth, land topology (mountains, hills, etc.), and permanent buildings. An effort of this type can be costly and requires the right detailed information as well as the maps and technical skills to carry out the investigation. It is common to use the services of a frequency coordination firm, such as Comsearch, for this effort. Part of the complexity is that the analysis must consider temporary propagation phenomena like rain scatter and ducting that would allow RFI to get from transmitter to receiver under other than standard clear-weather conditions.

Another TI concern is where an earth station is located near a high-power station such as an FM radio tower or L-band radar station. These frequencies fall within IF ranges of earth stations, making radiated or conducted interference a possibility. The energy can enter the IF chain of the earth station through connections between equipment or within the racks. This is another area where excellent shielding and grounding can provide a cure.

If the previous process indicates that some RFI may persist, then there is still a last search technique that involves on-site measurement of microwave signals. The approach involves using a standard receive antenna connected to an LNA and portable microwave spectrum analyzer, and literally sniffing for signals at the desired location. The process is simplified because the direction to the terrestrial transmitters is known from recorded data. However, it is not unusual for unrecorded transmitters to be uncovered. The potential for the earth station transmitter to cause RFI to the terrestrial station cannot be assessed so easily with field measurement (unless the earth station is already installed or a temporary transmitter can be put in place). For the receive RFI case, the test antenna should be raised to the same height as the installed earth station antenna. This in and of itself can be a challenge if the required antenna is of the 10-to-20-m variety. Otherwise, the potential for interference cannot be adequately measured at the particular location.

For larger earth stations, the objective of the RF clearance process is to prove that the operation of the RF terminal will not cause or receive unacceptable interference under any condition. The ideal situation is to obtain this type of clearance for the full frequency spectrum (typically 500 MHz for uplink as well as downlink) and the full GEO arc (or hemisphere, as appropriate). There will be cases when this is not possible, due to presence of strong emitters in particular directions that enter through the earth station main beam or primary sidelobes. This could be cleared through local shielding or agreeing not to point the earth station in particular directions.

References

[1]   Rees, Frank Jr., "Broadcast Facility Planning and Construction," *National Association of Broadcasters Engineering Handbook,* 9th ed., Washington, D.C.: National Association of Broadcasters, 1999, p. 1423.

[2]   Whitaker, Jerry C., "AC Power Conditioning," *National Association of Broadcasters Engineering Handbook,* 9th ed., p. 1427.

[3]   Lewis, Bernard T., and J. P. Marron, *Facilities and Plant Engineering Handbook,* New York: McGraw-Hill, 1973.

[4]   DeWitt, W. E., "Facility Grounding Practices," *National Association of Broadcasters Engineering Handbook,* 9th ed., p. 1445.

[5]   Braithwaite, I., "Lightning Protection of Electronic Equipment: An Arm's Length View," IEEE International Conference on Electrical Installation Engineering in Europe, Conference Publication no. 375, 8–9 June 1993, Institution of Electrical Engineers, London, p. 106.

[6]   Elbert, Bruce R., *Introduction to Satellite Communication,* 2d ed., Norwood, MA: Artech House, 1999, p. 223.

11

Principles of Effective Operations and Maintenance

Once the earth stations and user terminals are installed and activated, it becomes the responsibility of the operations and maintenance (O&M) organization to deliver the services and keep all ground systems in working order. In the analog past, most of the effort involved testing, tuning, repairing, and replacing components and subsystems; the conversion to digital technology eliminated much of the tuning, but made some operations more complex through new service options and an ability to add new users on the fly. The title of this chapter refers to *effective* O&M; it is our objective to effectively address the needs of the operating station. In this context, "better is the enemy of good enough." In other words, we need to work smart, doing the right job—not just doing the job right.

A general architecture for modern satellite network management is illustrated in Figure 11.1. From a user-terminal perspective, O&M is little more than applying power and making connections to end devices like personal computers, set-top boxes, display systems, and digital telephones. As the size and complexity of the ground segment grows, O&M begins to look more like telecommunications network management than equipment maintenance. The testing requirements have also become complex because the data must be encrypted, compressed, and packetized. Also, there is heavy

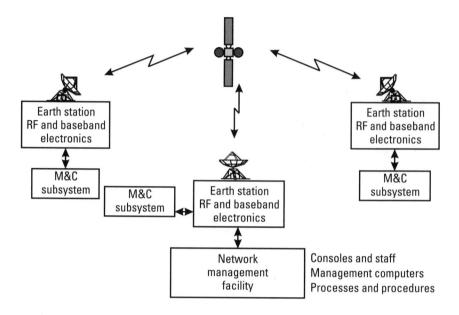

Figure 11.1 Modern satellite network management system.

emphasis on software to configure equipment, control data flow, and manage the overall network.

Earth station O&M is central to the proper functioning of the entire network. Like the human body, health of the whole depends on health of the constituent parts. Network well-being is tied to the well-being of the earth station (an operating organ), and the well-being of the earth station results from healthy components. The components covered in this chapter are addressed by the on-site operations and maintenance (O&M) staff. Operations people keep an active eye on things and notice problems when they are still potentially minor in nature. Maintenance staff may not have the benefit of direct observation and so depend on operations for first indications. There is another player that may or may not even be on scene—the engineering function, which designs and installs the elements and subsystems of the station. What we have is a partnership of these functions, each contributing to the overall success. However, this partnership does not happen automatically, as there is a normal tension between the three functions. Engineering takes requirements primarily from outside the station (e.g., from the business side) and provides a solution. The operations staff has to take over control of this solution and, as we indicate later, is as much a customer of engineering as is the business side. Maintenance sits between engineering and operations, seeing both points of view. The maintenance staff's technical competence will often be close to that of engineering, and so these staff members can (and do) take issue with the selected solution. They can also help operations understand the capability offered by the new system and adapt to it.

Table 11.1 provides a summary of the main functional areas of an earth station and the issues involved with providing proper O&M. These are guidelines for planning purposes and are not intended as maintenance practices. Specific requirements and specifications would be found within the O&M documentation published by equipment manufacturers and software providers. These must be expanded and further developed by O&M staff on an ongoing basis, based on experience with the equipment and services at the particular site. The purpose of this table is to identify the differing O&M needs across systems of a typical earth station. As a result, staff training and assignment tend to be appropriately specialized.

From a staff perspective, the expertise of engineers and operators at a major earth station correspond generally to the classes of equipment listed in Table 11.1. Training programs (both formal classroom/laboratory and on the job) are usually focused in the areas shown. Overall exposure to the earth station only comes on the job, either by coaching or through daily work. This is why new staff work as apprentices to the seasoned engineers and managers who have sufficient technical experience. The challenge is multiplied for any new installation because of limited exposure to the complete system and a lack of processes and procedures that mature and improve over time (which only happens if lessons learned are constantly identified and incorporated). Whether written or otherwise, these processes and procedures are essential to consistent and effective O&M of the site and the overall network (particularly if the earth station is a central communications hub and network management node).

Much of what is provided in this chapter is based on the personal experience and recent research by the author. In discussing this subject with real-world earth station engineers and O&M managers, we find consensus that earth stations are a demanding type of telecommunications facility. This is not difficult to understand—most large earth stations employ and depend upon systems of many types, along with the unique RF terminal and baseband equipment only employed in satellite applications. The people, services, and equipment that carry out the O&M function must act quickly and correctly in all situations that arise during expected and unexpected events, staying focused on required technical tasks throughout all manner of distractions. Once a network of users depends on a satellite network, any failure of service can take with it the very means of communication needed to

Table 11.1

Earth Station O&M Requirements by Functional Area or Subsystem

Equipment	Characteristics	Operation Requirements	Maintenance Requirements*
Antenna system	Fixed pointing	Check operation	Periodic inspection and protective care
	Tracking (limited motion, GEO)	Align on satellite and monitor performance; testing of tracking system	Periodic inspection, lubrication, and protective care (painting, replacement of flex waveguides, waveguide pressurization filter element changing)
	Tracking (full motion, GEO and non-GEO)	Configure for orbits (ephemeris), monitor tracking performance; testing of tracking system	Periodic inspection, lubrication, and protective care (painting, replacement of flex waveguides, waveguide pressurization filter element changing)
RF terminal electronics	Solid-state electronics	*Power On* status, gain and level verification	Periodic inspection
	TWT or klystron	High-voltage operation, redundancy switching	Performance measurement and periodic inspection; amplifier replacement
	IF chain	Proper configuration and function	Monitoring and periodic inspection and testing
Modem	Continuous, low and medium speed	Power on status; initial acquisition	Performance measurement and periodic inspection
	TDMA, medium and high speed	Power on status; acquisition; burst operation; BER	Performance measurement, periodic inspection and alignments
Baseband processing and multiplex	Constant bit rate, TDM	Configuration for services; monitor status	Periodic inspection; replacement of boards
	Variable bit rate, TDMA	Configuration for services; network operation; monitor status	Periodic inspection; replacement of boards
	Packet switching (routing)	Configuration for services; routing tables; monitor status, performance, and congestion	Periodic inspection; detailed load testing; replacement of boards
	Matrix switches	Identify failure modes and poor performance	Check full range of operations yearly

Table 11.1 (continued)

Equipment	Characteristics	Operation Requirements	Maintenance Requirements*
Computers and peripherals	Servers	Computer administration according to operating system requirements; configuration of databases and application programs per the operating need	Periodic inspection; replacement of boards; installation of software patches and upgrades, as required
	Workstations and monitors	User operation according to operating system requirements	Periodic inspection; software upgrades as required; attention to peripherals
Video monitoring	Video monitors	For continuously operating monitors, gradual degradation of quality	Periodic replacement
Monitor and control	Internal M&C system	Operation of earth station equipment from central location	Periodic inspection; software modification as required
	External M&C system	Operation of network of earth stations or user terminals	Similar to computer system maintenance
Facilities systems	Power systems	Monitor performance; checking and testing of UPS	Frequent inspection and load testing; replacement of filters, lubrication and other maintenance according to supplier recommendations; routine generator and fuel supply maintenance; UPS battery testing, servicing, and replacement
	HVAC	Monitor of HVAC system operation	Periodic inspections; alignments as required to provide temperature balance; filter replacement; lubrication
	Building	Monitor for leaks and other problems	Perform repairs as needed, immediately when needed

*These are general guidelines; appropriate manuals from equipment manufacturers should be consulted for specifics.

coordinate problem resolution, creating an additional imperative to keep all systems up.

11.1   Structure of Earth Station O&M

On any given day in the life of an earth station staff person, events are routine and probably nothing much happens. An exception is for video broadcast facilities where short-term program feeds are uplinked or downlinked on short notice. Even in the latter case, modern facilities are programmed to operate more or less automatically. The result is that experience in earth station O&M is rather like a soldier in ground combat—long periods of boredom punctuated by moments of stark terror (an exaggeration, of course, but to the point). In fact, a military organization provides an excellent example of how the people and equipment systems need to be organized.

11.1.1   Operations Organization

The operations portion refers to the human and machine aspects of controlling the equipment within the earth station and to which the earth station is connected (whether directly or over the satellite network). People involved with operations need to understand the functions associated with the systems, not necessarily how they work (and don't work). Good operations people can understand much about the systems they operate and manage, but their focus is on the services the earth station provides or controls. The maintenance portion, on the other hand, tends to be very technical in nature. These staff must understand how the various components are connected together, how they function individually and as a system, and the normal and abnormal behavior thereof. As an analogy in ground transportation systems, the bus is operated by a driver but maintained by the garage mechanic—jobs that are very different in terms of training and activity.

Operations covers most daily activities, like monitoring signals and channels, checking for continued performance from the RF equipment and antenna, and response to alarms. Maintenance is called upon when required. Management of this function requires a good understanding of the routine nature of the job along with its value to the enterprise. A high degree of esprit de corps is essential. In the end, service to users is mediated by the operations group, not by maintenance or even engineering. Engineering must sell off any new capability to operations. Being able to reject a project is one important power that operations must retain. Engineering has to have a strong hand, but the enterprise will not reach its goals (such as making money) without ongoing operations. However, operations staff need to be careful in applying this power. In summary, tension exists and there must be a balance of power among engineering, operations, and maintenance, and this power must be shared effectively and constructively.

11.1.2 Preventive Maintenance

One of the most important skills of O&M staff is to be able to identify and troubleshoot problems quickly, before they expand into service disruption or major failure of ground or even satellite equipment. Inexperienced and/or ill-equipped maintenance staff will identify the wrong source of the problem; in doing so, they can allow the problem to remain and even multiply. Some guidelines and organizational approaches to preventive and corrective maintenance are offered in the following subsections.

Modern telecommunications systems used to process video channels, switch telephone calls, and route information over the Internet are inherently reliable and maintainable through computerized control systems. In satellite communications, preventive maintenance tends to be as complex as the design of the systems themselves: different manufacturers have supplied hardware and software, and interfaces are often proprietary in nature. Similar requirements for user terminals, particularly VSATs, may be difficult because of the specialized nature of these devices coupled with the geographic area of the network. Success in this field demands a well-thought-out plan and a capable organization (which can use outside groups under contract). The plan should define responsibilities for maintenance functions, the specialized systems needed to facilitate O&M, and the processes to be used during routine activity (nonroutine activity is covered in the next section).

Table 11.2 provides a framework for developing a strategy for consistent and effective preventive maintenance of a major earth station facility (user terminals are discussed in Section 11.3). Even though the typical earth station is composed of equipment from several suppliers, we can assume that some type of centralized monitor and control (M&C) system is installed and functioning properly, usually operated by the operations side of the organization. The best time to install an M&C system is when the earth station is established in the first place. This can be done using an integrated approach (common in VSAT hubs and MSS gateway stations) or with a custom-designed system to support equipment supplied by different vendors. Older facilities can be fitted with a new M&C capability, provided that the original equipment can be operated remotely. This may still require a significant investment for special controllers, software, as well as a LAN to provide the necessary intercommunication.

Routine preventive maintenance is a team effort of the O&M staff (with engineering, marketing, and management playing a supportive role). By definition, the station is just one single point failure away from total customer dissatisfaction. Operations personnel can handle activities that are

Table 11.2
Preventive Maintenance Plan Structure for a Major Earth Station Facility Used as a VSAT Hub with
Internal Monitor and Control (M&C) System

Major Section	Type of Preventive Maintenance	Frequency*	Typical Activity
Outdoor RF equipment	Inspection and performance verification	Daily (Opns responsibility)	Correct beam alignment and gain
	Thorough test (requiring taking the antenna out of service)	Quarterly	Assess gain and polarization degradation; prevent moisture buildup leading to damage to waveguide
	Lubrication and surface restoral	Annual	Same as quarterly testing; detect change in mechanical and RF performance
	Monitoring of performance though M&C system	Continuous (Opns responsibility)	Detect any drop in transmit and receive power; frequency instability; monitor for loss of service
	HPA inspection	Monthly	Measure power output at unit; inspect high-voltage power supply and connections, front and back
	LNA, up-, and downconverter	Monthly	Inspect LNA and feed for rust, fungus, etc.; visual inspection of connections
	RF and IF backup equipment	Monthly	Verify function; check and replace filters and other consumables as needed
	Performance check: EIRP and G/T	Annual	Perform measurement of RFT characteristics using calibrated satellite carrier and/or test loop translator (TLT)
	Replace TWT	As and when required	Typically every three years, depending on loading and cycling
Baseband section	Monitoring of performance through network management or M&C	Continuous (Opns responsibility)	Measuring acquisition time, bit error rate (BER), lost frames and packets, time delay, dropped calls, etc.
	Detecting failed or degraded boards	As required	Based on internal fault monitoring and troubleshooting capability

Table 11.2 (continued)

Major Section	Type of Preventive Maintenance	Frequency*	Typical Activity
Computers and operating systems	Performance monitoring and load balancing	Continuous (Opns responsibility)	Based on internal CPU and LAN monitor and control system
	Routine maintenance of hardware and software	Periodic	According to manufacturer recommendations
Power and UPS	Monitoring of UPS, power and HVAC systems	Continuous (Opns responsibility)	Based on facility design, using centralized monitor system
	Routine maintenance of general plant and equipment	Periodic	According to manufacturers and special requirements of this location (e.g., tropical, arctic, or desert)

*These frequencies are for illustrative purposes, to gain an understanding of the relative timing of maintenance. Specific and appropriate values should be obtained from the relevant equipment suppliers.

performed more frequently. This includes routine measurements that can be performed remotely using standard measurement and test facilities, along with simple maintenance actions like replacing air filters and adjusting power levels. Maintenance engineers who have a detailed understanding of equipment failure modes and immediate corrective actions best perform tasks that require a significant effort or may involve taking something out of service. One thing that is of paramount importance is that all of the staff communicate with one another about what is or is not working correctly and any actions that were or should be taken. The best attitude to develop is that operations is the primary customer of maintenance (not only a source of irritation).

11.1.3 Corrective Maintenance

"Corrective maintenance" is a nice way of referring to the act of fixing things that are broken. In satellite communications, there are countless ways for failures to occur—we could say that Murphy's Law was really written around our industry. The good news is that the simplicity of satellite footprints and links can overcome the technical complexity of the components and systems that we need to meet service obligations. For example, our ability to monitor everything that is going on from one or a few locations provides the means to more easily identify and remove the cause in a short time frame. Our M&C capabilities are now so good that much of the corrective maintenance action is accomplished online, sometimes without human intervention (but always with humans being informed and involved).

It can be hard to define the requirements, processes, and skills needed for effective corrective maintenance. In virtually every real earth station facility this author has either managed or visited, the quality and experience of the staff were the most important success factors. Some years ago, COMSAT embarked on a project called the Unattended Earth Terminal (UET), the objective of which was to create earth stations and associated facilities that could be governed entirely through remote M&C capabilities. Efforts like this greatly improved the state of the art for centralized management and even corrective maintenance in some cases. However, experience continues to show that earth stations serving a large user community will require 24-hour on-site maintenance staff. The people who have this charge need the appropriate skill level and supporting test equipment to perform earth station maintenance in a professional and effective manner.

A suggested overall strategy for corrective maintenance is given in Table 11.3. Probably the most interesting and challenging aspect of this issue is problem source identification. Just how urgently one moves from examination to action depends on the degree of knowledge of the problem/solution as well as the seriousness of the service disruption itself.

The mechanical and microwave components of the antenna system are frequent causes of service difficulties. For this reason, there must always be someone either on site or readily available who knows the equipment and corrective actions that can be applied quickly. As potentially the weakest link in the system, the antenna deserves both the attention of the maintenance staff and the realization from management that the delivery of services is completely dependent on proper RF performance. Antennas with autotracking are a frequent cause of difficulty and for this reason the M&C system must include antenna operation as a central feature. Troubleshooting procedures are usually not offered by the antenna manufacturer, so the team must develop their own approaches for identifying and resolving problems. This depends on the experience base of staff coupled with what has been learned at the particular site. It is always beneficial to log corrective actions for review and inclusion in a procedures manual.

Table 11.3
Selected Corrective Maintenance Activities, Required Skill Levels, and Supporting Facilities

Symptoms	Possible Action*	Performed by	Support Equipment or System Required
Loss of RF link	Restore pointing toward GEO satellite (or tracking system operation for non-GEO constellation)	RF maintenance (electronic/ mechanical engineering)	Antenna control unit (ACU) adjustment; manual antenna mount adjustment; repair or replacement of motors, servo system or ACU
	Troubleshoot poor receive performance; identify failed unit, such as LNA or upconverter	RF maintenance (electronic engineering)	Detailed measurements of C/N using spectrum analyzer, RF power meter, frequency counter, etc.
	Loss of uplink signal, as measured at distant site; isolate down to the HPA and determine which component to adjust or replace	Network control and RF maintenance	Detailed measurement of C/N at distant site; use of local RF test equipment to verify loss of uplink power; failure of amplifier may require replacement of TWT
RF interference	Check polarization and adjust for minimum interference	Operations	Spectrum analyzer or power meter
	Check for adjacent satellite interference	Operations, in cooperation with network control	Spectrum analyzer
	Check for terrestrial interference	Operations and maintenance, supported by engineering	Spectrum analyzer and test antenna
Unstable service performance	Troubleshoot poor link BER or synchronization; replacement of circuit boards or entire unit	Baseband engineer or datacomm specialist	BER test set, spectrum analyzer, end user equipment
	Restore service to remote sites due to inadequate performance of link or end-user service; replacement of circuit boards or entire unit; remote device replacement from stock	Network control or management, aided by field repair staff and RF maintenance (if necessary)	BER test set, spectrum analyzer, end user equipment

Table 11.3 (continued)

Symptoms	Possible Action*	Performed by	Support Equipment or System Required
Failure of control functions	Conventional computer hardware troubleshooting; replacement of components or entire CPU; upgrading to latest software version if necessary	Computer support engineers; network administrator	Computer systems management interface; network analyzer
	Network management or M&C software failure	Network operations and computer support	Network management and computer systems management interface
Power disruption	Specialized troubleshooting and repair of UPS electronics, switching, battery systems, etc.	Facilities engineering, aided by UPS on call expertise	Power monitoring and management (particularly alarm systems)
Environmental failure	Identification of nature and source of difficulty (e.g., HVAC system)	Facilities engineering	Temperature and humidity monitoring system
	Building problems affecting operations or personnel safety	Facilities engineering	Water leakage, broken systems, cracks, etc.

*For example only; specific recommendations should be obtained for manufacturer or supplier documentation.

Outages may occur due to outdoor RF electronics attached to the antenna structure or located immediately adjacent to it. As we discussed in the last chapter, the people who take corrective actions must have safe and convenient access to high-power and high-voltage parts of the RFT. This author had the experience of having to work on the RF electronics within a poorly designed equipment container during a heavy rainstorm. The only solution was to quickly construct a crude tent over the container so that it could be opened safely. Working in pairs is another requirement when examining TWT and klystron HPAs, since the high voltages involved represent a serious danger. There was a death at an earth station where the unfortunate individual was working alone on an HPA.

Modem and baseband equipment are typically supported by the M&C and network management systems, allowing staff to identify the nature of the problem and correct it in an online fashion. On the other hand, the complex nature of the hardware and software within these systems sometimes makes the problem difficult to identify and resolve. The key here is to know the system intimately and to have at least some channels of redundancy to permit comparisons and changeouts as part of the troubleshooting process. The software and firmware in these systems quite frequently must be rebooted in the process, which can cause a problem in one area to propagate across a much larger range of services. The same can be said of the computers and their operating systems, which control and manage baseband equipment and user access.

Facilities, particularly electrical power and UPS, represent a potential weak link within the earth station service chain. This system is designed to provide reliable and stable power to the critical electronics throughout the facility; certain types of failures will be corrected automatically if the system is working correctly. Every facility that this author is aware of has had at least one failure of the combined power and UPS system, requiring some form of quick action. Often, this is due to a component failure, or an external cause, such as a very heavy rainstorm or fire. Sometimes it is human error during routine maintenance or installation of new equipment (including modifications to the power system itself). Such situations of increased risk should justify adding a temporary second level of backup, such as a mobile generator or alternate connection to the local power grid. Problems with or failure of the HVAC system (including the loss of power) produce discomfort for staff and a potential for equipment failure in time. These types of events may be uncontrollable and unpredictable, but corrective maintenance actions must be directed toward a solution.

Fairly routinely, the same staff involved with preventive maintenance can be called upon for corrective action when needed, since they are in relative proximity to problems when they occur. The trend toward fully integrated M&C systems reduces some skill levels for preventive maintenance, which may work against the corrective side of O&M. An approach to this problem might be to provide more in the way of active standby equipment and software, relying less on the staff to provide the direct "fix" to the problem. The cost of doing this must be traded off against that of hiring and training experts who can react like paramedics and doctors in the medical profession.

An important source of formal education and training is that provided by the equipment manufacturers for the RF and baseband portions of typical earth stations. These courses are directed toward the particular pieces of equipment, but vendors often put their devices in context in an overall network or earth station. The difficulty that management has is to select the most effective courses and provide them within an overall framework. In this manner, a particular course will provide the maximum benefit. For example, it makes no sense to send a new technician to a specialized course on klystron power amplifier maintenance until he or she has been introduced to the RF uplink configuration, types of amplifiers, and the principles of the gain budget.

Recruiting and training appropriate maintenance staff are decided challenges in our fast-moving industry. Training curricula that are specifically designed for satellite communications engineers and maintenance specialists are lacking. This is unlike the situation of standardized computer networking systems, like the highly recognized certificated programs of Novell, Microsoft, and Cisco Systems. For some years, UCLA Extension has provided more specialized sequences in satellite communications engineering, astronautical engineering, information systems, and the like. These training approaches are good in that they deliver technical content. However, what they cannot readily provide is actual field experience with real-world problems and issues that arise every day and every year for operators of satellite communications earth stations and networks. For this reason, it is always advisable to seek out competent outside technical resources among the equipment suppliers, system integrators, and consultants who address satellite communications.

11.2   Earth Station Alignment With the Satellite

Satellite transmissions can be received readily using the various forms of ground antennas discussed in previous chapters. This amounts to locating the satellite with respect to the point on the earth, pointing the antenna in the correct direction (assuming we are using a directional antenna), adjusting the pointing and polarization for maximum received signal strength and minimum interference, and adjusting the receiving chain for proper detection and operation of the service. When we are speaking of a transmitting earth station or user terminal, we must take many additional precautions to provide the proper signal through the satellite and to the receiving terminal, and protect the operation of the satellite and overall network from harmful RF interference. Experience has shown that the best way to accomplish this is through a centralized *network operation facility* that has the following capabilities:

- Can maintain direct communication and remain in contact with all transmitting earth stations, both at the time of initiation of service and throughout the service period. This can be done manually using human operators or in an online fashion over the network.

- Can monitor transmissions through the satellite to assure signal quality and to identify the possibility or existence of RF interference.

- Can control transmitting stations so that interference can be removed as quickly as possible. This control can be exercised manually by human communication or directly via a network. In the latter case, it is best if the network used for this control is separate from the satellite network being controlled (so that RF link loss or interference does not block central actions).

The type of facility that provides all of these functions is shown in Figure 11.2. We see here the consoles used to monitor signals over the satellite links, workstations with computers and telephones used for various forms of communication and analysis. Network control staff perform the real-time actions that maintain services and respond to situations involving RF interference and other problems that arise from time to time. In this world of

Figure 11.2 Network operations group of a satellite operator, the control point for access to satellite transponder capacity; it may also serve as the management point for part of the ground segment.

computer automation and instant communication, much activity can be performed by software and monitoring systems. The people, however, oversee this automation and are available at all times when the unexpected occurs.

Along with the network, major earth stations are under local human control as well. The procedure or protocol that they follow includes human conversation with network control, according to a script that is either written down or understood by virtue of experience with the overall operation. For example, when a transmitting TV earth station needs to bring up a carrier on the satellite, the network control function must first establish the authenticity of this operator. Then and only then can the process move through predetermined steps of qualifying the earth station and aligning the signals through the satellite repeater. Conducting this process for a two-way link adds the same steps for the second earth station (which is usually not physically located with the network control facility). These steps assure that the signals are on the right frequencies and within proper operating bounds in power and bandwidth. Then and only then can the earth stations perform end-to-end tests and commence revenue service. A detailed discussion of the organization and processes involved can be found in Chapter 11 of our previous work [1]; the following is a summary of the key points as they relate to maintaining earth station and network performance.

11.2.1 Prequalification of a Transmitting Earth Station

Prequalification is the process of measuring the specific technical parameters of an earth station prior to allowing access to the satellite or space segment. The idea is to provide the satellite operator with a high degree of assurance that this particular earth station will provide a satisfactory service without introducing more RF interference than allowed under existing operating rules and prior frequency coordination with other satellite systems. Traditionally, the long-established satellite operators like INTELSAT, EUTELSAT, and Inmarsat, have required very detailed tests that parallel the factory acceptance tests of the original equipment manufacturer. In time, testing cycles have been simplified to reduce the cost and delay associated with bringing an earth station into service. However, the fact remains that any earth station must operate correctly within its design limits. Background information can be obtained in reference texts [2].

Typical prequalification test criteria are:

- *Antenna pattern, sidelobe levels*—this guarantees that RF radiation outside the main beam is bounded correctly. Measurement of these parameters is restricted by the calibration of the test systems at the network control center. Often, the satellite operator will accept the pattern and sidelobe measurements from the antenna manufacturer, which were taken at the factory or even on a different test antenna of the same design. This may or may not be sufficiently accurate for the installed antenna, since the reflector and feed might not have been aligned correctly or there could be reflections from nearby structures and terrain features (e.g., hills or buildings).

- *Cross-polarization isolation*—this is critical to the operation of essentially every transmission over GEO satellites. With isolation values in the range of 25 to 45 dB (for relatively large reflector antennas), it can be difficult to measure the full range of values due to a particular ground antenna. This is complicated by the fact that the satellite itself may only perform to the 27 to 33 dB isolation level. Fortunately, a cross-polarization test of the main beam is relatively quick and should be performed both as part of prequalification and at the start of every transmission. Isolation in a linearly polarized system will change with time due to satellite motion and Faraday effect; hence, routine testing of cross-polarized interference is the norm. Circular polarization cannot be adjusted after installation, so any difficulty may require disassembly of the feed system.

- *EIRP*—this is determined by the transmit gain budget of the earth station (reviewed in Section 7.1), resulting in the uplink power radiated toward the satellite. During prequalification, the transmit power is measured at the input to the ground antenna and then combined by analysis with the gain of the antenna. Assuming that the antenna gain is known with reasonable accuracy (say 0.5 dB), the EIRP may be specified without an actual measurement in the far field of the antenna. The final measure would be over the operating satellite at the network control facility, using another known carrier as a reference. EIRP can be adjusted up or down using an attenuator located on the input side of the HPA and a power meter that is connected to the output side through a directional coupler.

- *G/T*—the ratio of receive antenna gain to system noise temperature, in dB/K, measured at the input to the receiving system. We assume that the factory-measured value of main beam receive gain is known ahead of time (this may not be feasible for antennas greater than about 9m in diameter). System noise temperature can be determined by analysis according to the procedure in Section 7.2. There are techniques for measuring system noise temperature directly, as indicated in Figure 11.3. In this approach, the received noise within a known bandwidth is compared with noise from a standard source. An attenuator permits the necessary adjustment of the standard to produce the final value. It is often best to measure G/T directly through the antenna using a calibrated carrier from the satellite as this is a system measure. This amounts to a backward link budget, where the measured C/N is used to calculate the unknown G/T, i.e.,

$$G/T = C/N - EIRP + A + 10 \log (BW) - 228.6, \, dB/K$$

Of course, this calculation is only as good as the accuracy of the individual terms. For example, we require a very accurate value of satellite EIRP (measured from another well-calibrated antenna system), and the total path loss, A, must be calculated precisely for the specific earth station location and satellite orbit position. The received carrier bandwidth, BW, should be a standard value using a calibrated filter on the spectrum analyzer of power meter. Another approach is to use a calibrated transmitting EIRP source located on a nearby mountaintop or tower (called a boresight). The difficulty is that the boresight is probably not located in the same direction as the satellite and thus there can be problems with terrain scatter. Accurate calibration data might be available from using a celestial microwave source such as the moon or a hot radio star (e.g., Cygnus A or Cassiopeia A) [3]. These methods offer potentially high accuracy but involve a considerable expenditure of time and money. For that reason, they are only appropriate where the antenna diameter is

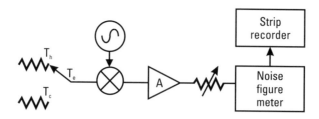

Figure 11.3 Measure of broadband RF noise using the Y-factor technique, where a comparison is made between the reference temperature source and the system under test.

large (in excess of 20m) for an earth station that must receive very weak signals. In general, it is most common to use a satellite as the reference as there are many to choose from. A well-calibrated earth station antenna to verify satellite EIRP provides the highest assurance of an accurate G/T calculation.

- *Antenna pattern characteristics*—main beamwidth, cross polarization, sidelobes, and backlobes. These are important characteristics for the operating system and service. If any of these measurements are not satisfactory, then it is likely that the antenna is defective or not installed correctly.

11.2.2   Uplink Access Test Prior to Service

An earth station that has passed its acceptance and prequalification tests is ready, in principle, to access the satellite for revenue-bearing services. The specifics of this procedure are normally defined and distributed ahead of time by the particular satellite operator. While there are no industry standard procedures, the process of the uplink access test (UAT) is fairly well understood by earth station operators around the world. Still, it would be appropriate to check with the operator of the desired satellite well ahead of time to assure that everything is in order. The following is a typical process for uplink access testing as would be used on a GEO satellite for the transmission of wideband video or data services. These tests would employ specialized test equipment from Agilent (and others), who also can supply integrated test systems under software control.

- *Uplink carrier check.* This is a quick test to verify that the EIRP, side-lobes, and cross-polarization isolation have not changed since preauthorization testing.

- *Carrier frequency.* It is necessary to determine the precise frequency of the carrier when unmodulated. This may or may not be the actual center frequency of the modulated carrier, depending primarily on whether it is a digital (e.g., PSK) signal or an analog (e.g., FM-TV) signal. Measurements made with a microwave frequency counter, while unmodulated, should be limited in time due to a high-power spectral density that poses a greater interference threat to other satellite networks. In most cases, center frequency can be measured with a digital spectrum analyzer.

- *Modulation bandwidth.* This important characteristic determines the transponder occupancy and is determined by the baseband bandwidth and the type of modulation. The satellite operator would have assigned a particular bandwidth and the earth station operator would ensure compliance with an IF or RF filter on the uplink side. In the case of digital (PSK) modulation in a saturated transponder, the bandwidth will likely spread due to sibeband regeneration by the satellite power amplifier, requiring careful adjustment of transponder backoff under direction from network operations. Bandwidth can be measured with a calibrated spectrum analyzer.

- *Burst and bit timing (if needed).* These are checks done for a digital satellite network using a particular carrier modulation, bit rate, channel coding, and synchronization scheme. For example, in a TDMA network, stations transmit bursts that must fall within nonoverlapping time slots. Receivers on the ground must be capable of synchronizing to the carrier, bit stream, and code sequence for the network to operate properly. This is performed automatically, both at initiation of service and again after any service interruption when a station must reenter the network.

- *RF channel amplitude response (if needed).* Depending on the type of carrier to be transmitted, it may be necessary or desirable to perform an IF or RF amplitude frequency response for the end-to-end link. Narrowband carriers (i.e., less than about 5 MHz in bandwidth) are normally not affected by earth station or transponder amplitude response. However, if the carrier bandwidth is 50% of the transponder bandwidth or greater, amplitude response can introduce baseband distortion. Carrier sidebands that are generated in the HPA of the earth station or satellite transponder would also need investigation to prevent out-of-band signal energy from overlapping into adjacent transponder channels. Typical equipment used to perform this test is illustrated in Figure 11.4.

- *RF channel group delay response (if needed).* Like amplitude response, this is a frequency-dependent characteristic that affects wideband signals. The measurement of group delay is more complicated than amplitude because of the need to track phase as the frequency is changed. The test setup, illustrated in Figure 11.5, would normally be part of an automated test system such as a network analyzer.

- *Threshold performance.* This is performed on the receiving side to verify that adequate link margin is provided. What we are looking

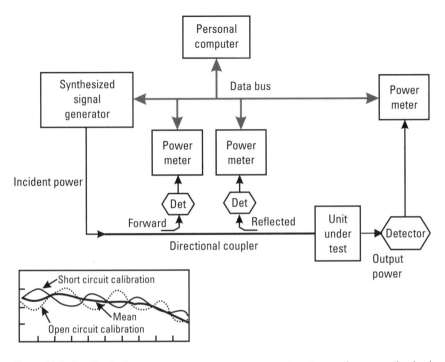

Figure 11.4 Amplitude-frequency response measurement system, using a synthesized signal generator as the input to the unit under test (UUT).

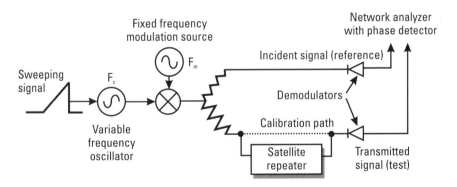

Figure 11.5 Dynamic measurement of group delay, using a modulated signal that counters the effects of frequency translation and Doppler due to a satellite repeater.

for is the value of clear-weather C/N and the margin that exists before a receiver/demodulator goes to threshold. The procedure can uncover a variety of potential problems, such as low satellite EIRP in a particular area of the footprint, excessive loss or high noise within the receiving system (internal or external to the receiving earth station), or an improperly designed or aligned demodulator.

- *Frequency stability and drift.* This is broken down into short-term and long-term frequency stability. Long-term stability is important to control the center frequency of the carrier to prevent adjacent channel interference and assure that ground receivers can maintain lock. A shift in center frequency in one direction over time is also referred to as frequency drift. Slowly varying frequency can potentially push the signal to the edges of the transmitting or receiving filters, producing amplitude or phase distortion. The effect of short-term stability depends on the associated time constants. If the frequency varies rapidly, then we can consider it as a form of phase noise that adds to other noise sources in the receiver.

- *Bit error rate performance.* Once the digital transmissions are synchronized correctly, the next step would be to measure the end-to-end bit error rate (BER). This can be performed using standard BER test equipment, a transmission impairments measurement set, or with capability built in to the modem or baseband equipment. The most common approach is to use a known bit sequence that is stored at the receiving end. The same sequence sent over the satellite link would include errors at the same rate as the user information. Then, the incoming sequence is compared with the locally stored value and the errors are counted as they are detected (by a simple comparison with what was sent). Alternatively, an estimate may be provided by monitoring the data check-sum bits of normal traffic flow.

11.2.3 Network Monitoring and Control During Service

An earth station that has been granted access to a satellite network is subject to continuous monitoring and control by the network operations facility. The processes depend on the transmission system design and approach to network intercommunication. The following are some top-level suggestions for conducting this aspect of the operation.

Network operations plays a critical role in real-time coordination or in case of RF interference or other difficulty. This role comprises:

- *Operator communication and control.* Earth stations and user terminals that can transmit to a satellite must be under the direct control of a network operations facility. As we mentioned previously, this can be through manual operators or via an automated network of some type. The requirements for communication and control would be defined in a user agreement or contract, making these provisions binding on all stations that can transmit. The specific requirements will differ from application to application and system to system, but the need is nevertheless the same.

- *Contact information.* All stations must have a direct communication link with network operations. In an automated system, this type of data can be sent over the satellite link during station acquisition. Location data (if required) can be derived from GPS measurements or preset for the particular latitude and longitude. For manual operations, the information would be transmitted via voice or a data signal.

The concepts and operating processes discussed in this section are limited to the introductory level. In a real-world situation, these would be embodied in plans and procedures that are maintained by the network control facility and shared by each accessing earth station. Manually controlled satellite networks used for TV transmission follow these guidelines directly. For automated networks that provide online services to VSATs and mobile user terminals, the procedures are contained in thousands of lines of computer code. Thus, it can be a complex matter to identify and eliminate problems that are caused by the control system itself. No matter what happens, operators and earth stations and networks need to understand that they are using RF propagation over long distances, and that all frequencies are shared by a wide range of users. For this reason, a good attitude toward all users of the spectrum is the best approach for minimizing incidents of actual RF interference and providing the highest assurance of quality services via satellite.

11.3 Service Support for User Terminals

Delivery of services by user terminals capable of transmitting signals to satellites introduces several complexities to the overall operation and maintenance process. Users include purchasers of VSAT networks, satellite mobile devices (such as mobile phones), and specialized two-way terminals used for video-conferencing or distance learning. We must assume that the user terminal, while being capable of transmitting, is not directly operated by users (other than aligning the antenna and turning the device on or off). In fact, communication with the user would not be possible until the terminal is working correctly within the overall network. Loss of direct communication should result in automatic disabling of uplink transmission to the satellite.

The best parallel to this is a land-based mobile phone that automatically registers itself with the local cellular or PCS network when it is first activated. When the mobile phone is turned on, it seeks the strongest control carrier within its operating bandwidth and then transmits a request for entry into the network. Subsequently, the cellular network accepts the station and allows it to place calls on demand from the user. The user is notified that service is ready by an appropriate readout on the LCD screen of the mobile phone. Similarly, a satellite network of user terminals must be capable of this type of automated alignment when the user turns on the device. For example, a user who has just turned on a terminal and aligned it with the satellite should not have to perform many complex functions to initiate communication; contact should be achieved simply by making a phone call to a customer support center, for example. The satellite network should be capable of initiating transmission from the terminal and incorporating it into the network. Of course, this presents more potential for RF interference, due to the wider coverage of a satellite as compared with a given cellular base station. The network control facility still provides that single point of service monitoring and control, although the vast majority of activity is fully automated using software that runs in the network and the terminal devices. As long as user terminals and hub earth stations behave correctly, there is no need for intervention by human operators at the network control facility.

11.4   Management of O&M

Recognizing that it would be difficult to generalize with respect to O&M management, we can take a rather positive and proactive view of the task. The various functions within O&M are vital to the system and deserve the best from everyone. Management must be able to play the game along with the working operators and maintainers. This is a place for MBWA (management by walking around). You can learn a lot by observing and asking simple questions.

Here are some points learned on the job over the last several decades:

- People are the solution, not the problem. When things go wrong, don't automatically assume that someone has made a "mistake." Most employees do their jobs according to what they understand the requirements to be. Failures involving people are more often than not failures of procedures and management.

- Manage with loose reins. This philosophy (as well as the next one) was promoted by the late Col. John Sedano (USAF, retired), former vice president of operations at Hughes Communications. It is clear that management in O&M has to be very active and involved. At the same time, people should be allowed to do their jobs without too many strings attached. See also the last item in this list.

- Anticipate Murphy. Expect the unexpected, and plan accordingly.

- Believe the data. Any indication of difficulty, no matter how small, should be taken seriously. This is the approach taken by the U.S. Navy in its nuclear submarines. Likewise, believe what someone on the team (or a customer) is telling you. Do the homework to check it out.

- Include backup, particularly if you don't have to pay extra for it ("Elbert's First Rule"). Backup and alternate paths are often expensive when purchased as such. Better to include it with the initial system when the cost is low. Also, if there is a working capability in place when a new one is being introduced, think about keeping it running as a low-cost backup.

- Don't pass the buck (or expect someone else to take care of the problem). As managers, the task and problem are yours. With a competent staff, any problem can be solved. It may be a matter of time, so don't rush the cure.

- Check your end first. There is a tendency to point fingers at the other end or at the satellite. But it is worth the time and effort to be sure that your own house is in order.

- Choose the best people. Use a thorough evaluation process, such as that described by Bradford Smart [4].

- Put the right person in the right job, and vice versa. In terms of our personal contributions, we are not all created equal—each person has a unique ability, talent, and interest. As managers, we need to identify these strengths and help others to reach their potential.

Also, motivation is in there for all of us, but we are not motivated by the same package. Technical staff are perhaps a bit unique in this regard, as discussed by Douglas Soat [5].

- Build a culture of innovation, integrity, and customer focus. We are in a New Economy, where telecommunications and the Internet have changed all the rules. But have they? In a technical organization, it is important that people feel that they and their contributions are valued, and that the organization counts for something. We can all take a lesson from Charles Schwab, now the leading online/offline stock brokerage, on how important culture still is [6].

- Lastly, shift the load of responsibility to the employee. According to Chuck Hollins, supervisor of Los Angeles County's extensive wide area network, if we give our staff the tools they need, we can expect them to shoulder the responsibility. In this way, we drive accountability directly to the user interface so problems are resolved quickly.

11.5   A Final Word

This book is perhaps innovative in its cohesive treatment of ground segments for satellite communications. We have tried to put in one place as much information as our experience suggests. Clearly, there is a lot more that can be said and explored. The nature of the ground segment and earth stations is constantly changing, as new applications appear alongside those that have been around for 30 years. Certain aspects never change—we still speak of the uplink and the downlink, and information must be processed and distributed through the area coverage capabilities of satellites. We hope that this effort has helped reduce the obstacles for those who want to apply ground segment technology to practical communications problems on earth.

References

[1]   Elbert, Bruce R., *Introduction to Satellite Communication*, 2d ed., Norwood, MA: Artech House, 1999.

[2]   Lance, Algie L., "Microwave Measurements," Chapter 9, *Handbook of Microwave and Optical Components*, Vol. 1, *Microwave Passive and Active Components*, ed. Kai Chang, New York: Wiley, 1989, p. 419.

[3]   Kraus, John D., "Radio-Telescope Antennas," Chapter 41, *Antenna Engineering Handbook*, 3d ed., ed. Richard C. Johnson, New York: McGraw-Hill, 1993, p. 41.

[4]   Smart, Bradford D., *Topgrading: How Leading Companies Win by Hiring, Coaching and Keeping the Best People*, Paramus, N.J.: Prentice Hall, 1999.

[5]   Soat, Douglas M., *Managing Engineers and Technical Employees: How to Attract, Motivate and Retain Excellent People*, Norwood, MA: Artech House, 1996.

[6]   Pottruck, David S., and Terry Pearce, *Clicks and Mortar: Passion Driven Growth in an Internet Driven World*, San Francisco: Jossey-Bass, 2000.

About the Author

Bruce R. Elbert, author of seven Artech House books on satellite communications and information technology, has 35 years of technical and management experience in these fields. As managing director of Application Strategy, a consulting and training firm based in California, he assists corporations and government agencies in developing their technical and operational capabilities in satellite and broadband communications systems. Mr. Elbert was with Hughes Electronics for 25 years. He held several key management positions, including director of marketing and engineering for the Galaxy satellite system, and senior communications engineering manager for Indonesia's first domestic satellite network, Palapa A. His first industry position was as a satellite systems engineer with COMSAT back in the early days of communications satellites. He holds masters' degrees in business administration (Pepperdine University, 1985) and electronics engineering and computer science (University of Maryland, 1972), and obtained his bachelor of electrical engineering from CCNY in 1965. Mr. Elbert served in the U.S. Army as a captain and radio communications officer, including a tour in Southeast Asia with the 4th Infantry Division. He is series editor of two Artech House Libraries: *Space Technology and Applications* and *Technology Management and Professional Development.*

Index

**Recent Titles in the Artech House
Space Technology and Applications Series**

Bruce R. Elbert, Series Editor

Introduction to Satellite Communication, Second Edition,
Bruce R. Elbert

Low Earth Orbital Satellites for Personal Communication Networks,
Abbas Jamalipour

Mobile Satellite Communications, Shingo Ohmori

The Satellite Communication Applications Handbook,
Bruce R. Elbert

*The Satellite Communication Ground Segment and Earth Station
Handbook,* Bruce R. Elbert

Understanding GPS, Elliott D. Kaplan, editor

*Gigahertz and Terahertz Technologies for Broadband
Communications*, Terry Edwards

For further information on these and other Artech House titles,
including previously considered out-of-print books now available
through our In-Print-Forever® (IPF®) program, contact:

Artech House
685 Canton Street
Norwood, MA 02062
Phone: 781-769-9750
Fax: 781-769-6334
e-mail: artech@artechhouse.com

Artech House
46 Gillingham Street
London SW1V 1AH UK
Phone: +44 (0)171-973-8077
Fax: +44 (0)171-630-0166
e-mail: artech-uk@artechhouse.com

Find us on the World Wide Web at:
www.artechhouse.com